Mind, Brain, Quantum AI, and the Multiverse

Mind, Brain, Quantum AI, and the Multiverse

Andreas Wichert

CRC Press
Taylor & Francis Group
Boca Raton London New York

CRC Press is an imprint of the
Taylor & Francis Group, an **informa** business

A CHAPMAN & HALL BOOK

First edition published 2023
by CRC Press
6000 Broken Sound Parkway NW, Suite 300, Boca Raton, FL 33487-2742

and by CRC Press
4 Park Square, Milton Park, Abingdon, Oxon, OX14 4RN

CRC Press is an imprint of Taylor & Francis Group, LLC

ISBN: 978-1-032-14960-8 (hbk)
ISBN: 978-1-032-15532-6 (pbk)
ISBN: 978-1-003-24454-7 (ebk)

DOI: 10.1201/9781003244547

Typeset in Latin Modern
by KnowledgeWorks Global Ltd.

Publisher's note: This book has been prepared from camera-ready copy provided by the authors.

For my son André and my wife Manuela

Contents

Preface

There is a long-lasting controversy concerning our mind and consciousness. The book proposes a connection between the mind, the brain and the multiverse. We introduce the main philosophical ideas concerning mind and freedom and explain the basic principles of computer science, artificial intelligence of brain research, quantum physics and quantum artificial intelligence. We indicate how we can provide an answer to the problem of the mind and consciousness by describing the nature of the physical world. Our proposed explanation includes the Everett Many-Worlds theory. The book tries to avoid any nonessential metaphysical speculations.

Introduction - Chapter 1

The first chapter deals with the mind from the perspective of philosophy. Plato introduces the allegory of the cave. Reality is not what it seems. Plato did not trust our human senses: What is reality and what is imagination? The allegory describes how people who depend only on their senses are analogous to prisoners who spend their lives in caves, looking at shadows of reality. Plato introduced the distinction between the physical and eternal worlds. For Plato, the soul and the body are of different worlds. Aristotle thinks like his teacher that the soul is a part of a living being and that it is distinct from the body. He does not believe in a Platonic eternal world of abstract objects. René Descartes, a dualist, assumes that our bodies and animals are machines, and the mind is nonphysical. The mind influences bodies and pilots the machinery. For the modern philosopher Gilbert, the mind is an illusion resulting from logical errors of thought. Cartesian theory holds that mental acts determine physical acts. This theory, according to Ryle, is "the myth of the ghost in the machine," there is no separation of mind and matter. Computationalism states that the human mind is an information processing system, and that cognition and consciousness together are a form of computation. The mind is a computational system that is realized by the brain. Free will is an illusion.

Computer Metaphor - Chapter 2

The early history of algorithms is introduced followed by Gödel's incompletess theorem. The Entscheidungsproblem is presented, and the Turing machine is introduced. The universal Turing machine is an abstract model of a computer. Computational complexity theory addresses questions regarding which problems can be solved in a finite amount of time on a computer. The Church-Turing thesis states that any algorithmic process can be simulated on a Turing machine. Two classes of practical computers are presented: Analog and digital computers. The history of artificial intelligence is presented and the theory of mind. A complex system is a system composed of many parts, which interact with each other. An example of a complex system is the human brain where many neurons interact with each other. One example of a complex system whose emergent properties have been studied extensively is the cellular automata. Marvin Minsky proposed the society of mind theory in which a vast society of individually simple processes known as agents interact with each other.

Brain - Chapter 3

Neurons are the building blocks of the brain. The memory is stored in the synapses. The human nervous system can be divided into the central nervous system that corresponds to the brain and the spinal cord. The human visual system is the most well-understood system of the human brain and can be seen as a computer metaphor. The neural assembly theory was introduced by Donald Hebb. He proposed a connection between the structures found in the nervous system and those involved in high-level cognition such as problem solving. It is unclear what adaptive advantage consciousness could provide at all. Global workspace theory (GWT) is a simple cognitive architecture that can be explained in terms of a "theater metaphor." In the metaphor, the stage corresponds to the contents of consciousness, like actors on the stage. In the 1980s, Benjamin Libet asked subjects to choose a random moment to flick their wrists while he measured the associated activity in their brains and concluded that we have no free will in the initiation of our movements. The results suggest that decisions made by a subject are first being made on a subconscious level before one is consciously aware of having made it. António Damásio argues that consciousness provides two functions: To construct narratives and to feel one's emotional response to them. They give humans and other conscious animals the ability to imagine possibilities, evaluate them, and to plan future actions. The explanation hypothesis assumes that the identity is more related to the human senses than it is to an algorithmic device.

Quantum Reality - Chapter 4

Information is a part of the physical universe as matter and energy. Today, first small quantum computers begin to appear. A quantum computer represents information by qubits. In quantum computation, there are two principles (algorithms) that speed up the computation: Fourier Transform and Grover's amplification algorithm. Quantum artificial intelligence is introduced by linear algebra-based quantum machine learning and symbolical quantum artificial intelligence. The initial motivation for relating quantum theory to consciousness was related to free will in a deterministic world, since the quantum world is nondeterministic. The dualism of mind and matter involves the unresolved problem of how the mind and brain could interact. The relationship of union between the mind and the body is resolved by quantum measurement. Sir Roger Penrose argues that elementary acts of consciousness cannot be described by a Turing machine. Such noncountable elementary acts of consciousness are only present in the wave function collapse (measurement). Penrose assumes that the wave function collapse is not random; it represents an objective reduction.

Multiverse - Chapter 5

The modern version of the wave-function collapse in quantum mechanics is based on decoherence and leads to the multiverse interpretation of quantum mechanics. The decoherence causes the universe to develop an emergent branching structure. One can interpret the branching structure either as a tree structure or as parallel different histories. Time is not continuous since there are no measurable continuous in quantum physics. The notion of continuous range of possible values of time is just an idealization. The same goes for the direction of time that is a prior assumption based on the branching structure of the multiverse. The multiverse does not exist in external space nor time since there is nothing more beside it. The observer's mind should not be in a superposition with himself. Because of this, in the many-minds interpretations it is argued that the distinction between worlds should be made at the level of the individual observer. Human bodies are in a superposition, but minds have definite states that are never in superposition. All possible lives, personalities that do not violate the laws of physics, are facts since they represent a valid history in the multiverse. We are one of this possible personality. Our choices are defined by our personality explained to us by our minds, our consciousness is the unchangeable part of us. It is nonphysical and it senses the world. It does not interact with the brain in the other way beside it.

Conclusion - Chapter 6

We analyze the different assumptions of the mind from an ethical perspective. How do I know that other beings with minds exist? All others apart from me are possibly automata, zombies or non-feeling individuals who behave similarly to me. Robots can exhibit high levels of cognition, but cognition and mind are different terms. Cognition is closely related to human intelligence and can be simulated by computers and robots. On the other hand, mind is vaguely defined involving consciousness, a combination of cognition and emotion, and unconscious cognitive processes. Earth has minds other than ours: Those of the animals. Does our universe have any other civilizations? We cannot communicate efficiently with animals. We are not interested in doing so. How then can we expect to communicate with other life forms on other planets? We will only be able to communicate with technical civilizations such as ours. *Summa Technologiae* is a 1964 book by the Polish author Stanisław Lem. The book tries to address future problems. Despite its age, the book has lost none of its relevance. We examine some of the problems described in the book in the context of our findings from the subsections on other civilizations. Technological evolution and science will give us increasing abilities. However, science alone cannot answer some questions, such as those about morality and ethics, love, the meaning of our lives or the creation of beauty.

This book is an essential compilation of knowledge in philosophy, computer science, biology and quantum physics. It is written for readers without any requirements in mathematics, physics or computer science.

Author

Andreas Wichert studied computer science at the University of Saarland. He graduated in 1993. He studied philosophy and computer science at the University Ulm. In 1999, he earned a bachelor's in philosophy and a PhD in computer science in 2000. Since 2006, he has been an assistant professor in the Department of Computer Science and Engineering, University of Lisbon. His lectures are on machine learning and quantum computation. His research focuses on neuronal networks, cognitive systems and quantum computation. He has published four books and over 100 scientific articles. In addition to science and philosophy, he is a passionate artist with several exhibitions in Germany, Poland and Portugal.

CHAPTER 1

Introduction

T HE MIND is a tricky topic leading to difficult questions that we will try to answer, such as what is the relation between the mind and the body, or, generally, what is the relation between the mental and the physical properties? How does the identity of a mind arise in time? What does it mean for someone to be him or herself across two moments in time? By looking on an individual's behavior, one can tell if this person is experiencing pain, but only the individual feels it. Physical properties are equally observable by any person, but the mental consciousness of such events is private. Why does this phenomenon occur?

Prior to the development of modern science, all such questions were addressed by philosophy. Metaphysics denotes philosophical enquiry by a nonempirical scientific method. The word "metaphysics" is derived from two Greek words that together mean "after the natural." The existence of mind in a world composed of matter is one of the greatest unsolved metaphysical problems. When dealing with the problem of mind, science is often unable to provide answers or the metaphysical explanations are hidden behind scientific language. The reader must distinguish between a scientific empirical explanation and a metaphysical hypothesis that is not based on the scientific method.

1.1   MIND AND DUALISM

1.1.1   Plato

We begin our investigation of the mind and the brain with a philosophical enquiry. We start with one of the most important and influential individuals in Western human history. Plato (428/427 or 424/423–348/347 BC) was an Athenian philosopher during the classical period in Ancient Greece, along with his teacher Socrates and his most famous student Aristotle. In the popular film The Matrix, the main character is trapped in a false reality created by a computer program. The program was created by machines that took over the planet. The story was inspired by Plato's allegory of the cave. Reality is not what it seems. Plato did not trust our human senses: What is reality and what is imagination? When we dream and hallucinate, we confuse reality with our imagination. Often, we are not aware that we make mistakes and that things are not what they seem. To discover the real nature of the world, according to Plato,

DOI: 10.1201/9781003244547-1

Figure 1.1 The allegory of the cave. The prisoners are chained in the cave and are unable to turn their heads; they cannot see that the shadows are generated by puppets that are held by puppeteers.

we must use our rational minds. Plato introduces the allegory of the cave in his book "The Republic" [216]. The allegory describes how people that depend only on their senses are analogous to prisoners who spend their lives in caves, looking at shadows of reality. In the cave, the prisoners are chained and cannot turn their heads, so they cannot see that the shadows are generated by puppets that are held by puppeteers. They confuse the shadows with real objects. They cannot find the real causes of the shadows. They never know that the shadows are not real objects but puppet shadows on the wall in the cave, which is lit by a fire (see Figure 1.1).

If a prisoner could escape and discover that a whole world existed outside the cave, none of the other prisoners in the cave would believe him or her.

How can we attain freedom? According to Plato, we should use our rational minds, which will lead us to the theory of ideas (or forms). This theory denies the reality of the material world, considering it only an image or copy of the real world.

Plato assumed the existence of three worlds: The world of material objects, the world of our mental images as grasped by our senses and the world of unchanging eternal abstract objects. The mental representation of objects changes constantly; it is nothing more than a story to which we listen. However, eternal objects do not change and can only be grasped by pure reason. The eternal world exists outside time and space. An example of an eternal world is mathematics. The distinction between the physical and eternal worlds is introduced in the book "Timaeus" [217]. The physical one is the world, which changes and perishes; the eternal one never changes and can be apprehended only by reason. Plato introduced the concept of the immortal soul; he assumed that the intellect is immaterial because it interacts with eternal objects. The soul strives to leave the body in which it is imprisoned. Plato was a dualist; the soul and the body are of different worlds. However, the interaction of the body with the soul remains a deep mystery since no explanation is given about how the two worlds interact.

1.1.2 Aristotle

Plato's famous student, Aristotle (384–322 BC) thinks like his teacher that the soul is a part of a living being and that it is distinct from the body [10]. He does not believe in Platonic eternal world of abstract objects; he does not assume that the soul can exist without the body. The soul is a nature of a human being, a property of the human body. It is defined by its relationship to an organic structure of a living being. Animals and plants have souls, soul is a body that has life. There is a hierarchical arrangement of multiple souls, the vegetative soul, the sensitive soul, and the rational soul. The vegetative soul can grow and nourish, the sensitive soul can experience sensations and move. The unique part of the human, the rational soul knows and understands. This soul is called the mind, it can reason and plan the courses of action. Besides the senses, such as color, sound, taste and smell, there are the inner senses such as imagination and memory. Aristotle does not support the Platonic dualism. He also does not support the materialism of the early Greek philosophy, the so-called pre-Socratic philosophy. This is the philosophy before Socrates and Plato [208], as advocated by Parmenides (530–460 BC) or Democritus (460–370 BC) [274]. Materialism holds that matter is the fundamental substance in nature, and that all things, including mental states and consciousness, are results of material interactions. Aristoteles seeks a middle course between two extreme viewpoints, materialism and dualism, by pointing out that these are not the only exhaustive options.

1.1.3 Neoplatonism

Neoplatonism is a strand of Platonic philosophy that emerged in the second century AD [290], the most known philosopher of this direction is Plotinus (204/5–271 AD). Simplified the Neoplatonism states that there is an eternal immortal world-soul that is part of our mind. The immortal-world soul does not change and is part of our mind that feels. There are no individual immortal souls that are trapped the living body as stated by Plato. Something eternal is always there timeless and cannot change. That is in contradiction to the physical world that is present in time. Consciousness is not an emergent property of the material world, but the part of the immortal-world soul.

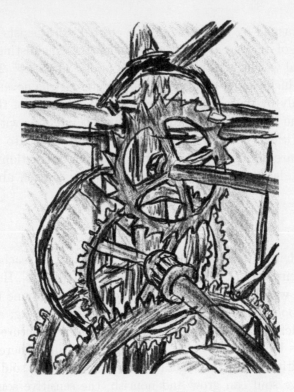

Figure 1.2 Parts of a mechanical clock machine from the 16th century, the convent of Christ in Tomar, Portugal.

1.1.4 Descartes

Between 1280 and 1320, a true mechanical clock mechanisms have been developed that were installed in the church towers (see Figure 1.2).

In the 15th century spring-driven clocks appeared, during the 15th and 16th centuries, clockmaking flourished, and the first automaton arrived like the clockwork monk that could walk, strike his chest and roll his eyes (see Figure 1.3).

René Descartes was inspired by this kind of automaton and began to investigate how the mind interacts with the body [80], see Figure 1.4. For Descartes, our bodies and animals are machines that work according to their own laws like the mechanical automaton. The mechanical automaton become the metaphor for living bodies. Bodies are complex machines, they could be replaced with cogs, pistons and cams. For Descartes the mind is a nonphysical and nonspatial. He identifies the mind with consciousness and self-awareness. The universe contains two radically different kinds of substances, the nonspatial mind that is thinking, and the physical body that is not thinking. The mind influences bodies and pilots the machinery. Like in a "Super Mario" video game where we pilot our virtual "Mario" through the artificial world full of dangers. According to Descartes, the mind and the body are connected through the pineal gland in the human brain since this part of the brain is not duplicated. The main problem of how the interaction takes pace remains unsolved. How can a nonphysical and nonspatial mind interact with the physical world? In the causation,

Figure 1.3 A clockwork monk that could walk, strike his chest and roll his eyes, South Germany or Spain ca. 1560.

two billiard balls influence each other, mental acts should influence the physical acts in the same manner as the billiard ball influence each other. The concept of human beings, according to Descartes, is a composite entity of mind and body. For Descartes mind can exist without the body, but the body cannot exist without the mind. Mind and body are closely joined, but what exactly is the relationship of union between the mind and the body of a person is an unsolved mystery.

1.2 PHILOSOPHY OF MIND

1.2.1 Birth of Modern Science

Sir Isaac Newton (1642–1726/27) formulated in the book "Principia" [203] the laws of motion and universal gravitation. With the mathematical description of gravity, Kepler's laws of planetary motion could be derived and the trajectories of comets could be predicted. The motion of objects on Earth and celestial bodies could be accounted for by the same principles.

In the 16th century, Francis Bacon rejected scholastic metaphysics, and argued strongly for empiricism [19]. How can we get free from the cave? According to Plato, we should use our rational mind. According to Francis Bacon, our rational mind can folly us as well. To understand the nature of the reality, we should execute empirical experiments. This is a radical change that transformed the natural philosophy into an empirical activity deriving from experiment, unlike the rest of philosophy. The

Figure 1.4 René Descartes (1596–1650) was a French philosopher, mathematician, scientist and lay Catholic who invented analytic geometry, linking the previously separate fields of geometry and algebra. One of the most notable intellectual figures of the Dutch Golden Age, Descartes is also widely regarded as one of the founders of modern philosophy and algebraic geometry.

scientific method and empirical science itself were born with Francis Bacon being their father. Science and philosophy have been considered separated disciplines ever since. In science, knowledge is based on experience that is subject to continued revision.

1.2.2 Gilbert Ryle

In the book "The Concept of Mind" [238] the philosopher Gilbert Ryle (see Figure 1.5) assumes that the mind is a an illusion resulting from logical errors of thought. The book is the founding document of the philosophy of mind that denies the Cartesian dualism of Descartes. Ryle states that the doctrine of body/mind dualism was the official doctrine that is not valid anymore. Ryle rejects Descartes' theory on the grounds that it assumes that the mental processes could be isolated from physical processes. Mental processes are instead merely intelligent acts that may not necessarily be produced by highly theoretical reasoning or by complex sequences of intellectual operations. Mental processes may be explained by examining the rules that govern those actions.

Cartesian theory holds that mental acts determine physical acts. This theory, according to Ryle, is "the myth of the ghost in the machine," there is no separation of mind and matter. There is no contradiction between saying that an action is

Figure 1.5 The philosopher Gilbert Ryle (1900–1976). He wrote the book *The Concept of Mind*, a founding document of the philosophy of mind.

governed by physical laws and saying that the same action is governed by principles of reasoning. Ryle put the final nail in the coffin of Cartesian dualism and replaced it with philosophical behaviorism.

1.2.3 Daniel Dennett

The book Consciousness Explained written by the American philosopher Daniel Dennett 1991 [79] is in the tradition of Gilbert Ryle. Dennett describes consciousness as an account of the various calculations occurring in the brain close to the same time. He assumes that we humans are machines. Because of this, in near feature artificial machine will have consciousness as well. Dennett states that the best reason for believing that robots might some day become conscious is that we humans are a kind of machines and are conscious. Dualism is old fashion stuff, still popular since some humans assume that mind should be protected from the methods of the physical sciences. The modern dogma states that mind is a process, our illusion is that we assume that it is more than a process. The conscious, our self-sense, is the content of a model that is created by our brain, the self-sense is an illusion.

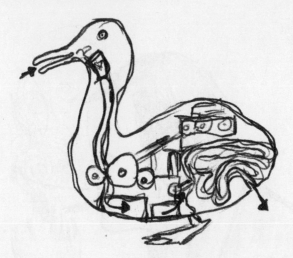

Figure 1.6 The Digesting Duck was an automaton in the form of a duck created by Jacques de Vaucanson. The duck ate kernels of grain and defecated them. It was an illusion, since the food was collected in one inner container and the pre-stored feces was produced from a second so that no actual digestion took place.

1.3 COMPUTATIONALISM

1.3.1 Automaton and Calculators

Digesting Duck was an automaton in the form of a duck created by Jacques de Vaucanson and unveiled on 30 May 1739 in France. The duck eats kernels of grain and defecate them. It was an illusion, since the food was collected in one inner container, and the pre-stored feces were produced from a second, so that no actual digestion took place (see Figure 1.6). However, Jacques de Vaucanson hoped that a truly digesting automaton could one day be designed and become a true metaphor for an animal body. Around the same time simple calculation processes, which usually required human mind were mechanized. In 1642, Blaise Pascal invented the mechanical calculator called Pascaline. The comptometer, introduced in 1887, was the first machine to use a keyboard that consisted of columns of nine keys (for each digit). Charles Babbage (1791–1871) sketched the first mechanical machine (a Difference Engine) for the calculation of certain values of polynomial functions [140]. In 1822, Charles Babbage presented a small cogwheel assembly that demonstrated the operation principles of his Difference Engine, a mechanical calculator, which would be capable of holding and manipulating seven numbers of 31 decimal digits each. It could automatically compute and print mathematical tables. It was never built in Babbage's lifetime. His design was only constructed in 1989–1991, by the Science Museum, London (see Figure 1.7). The main obstacle in Babbage times was that no norms existed, for example, for the size of screws. Each producer had its own standard making it nearly impossible to construct complex machines such as the Difference Engine. After the failure of manufacturing the Difference Engine, Babbage worked to even design a more complex machine. With the goal of mechanizing calculation steps, Babbage sketched the first model of a mechanical universal computer and called it the Analytical Engine.

Figure 1.7 The Difference Engine number 2, constructed in 1989–1991, by the Science Museum, London.

Luigi Federico Menabre, a military engineer, made notes and, subsequently, with the help of Babbage, published a paper in French on the device. Ada Lovelace (1815–1852) translated the paper it into English. Babbage had specified the instructions that the Analytical Engine would perform [140]. This allowed Babbage and Lovelace to write programs for the nonexistent computer. Babbage was never able to build the Analytical Engine. Ada Lovelance thought about the computing power of such a machine. She argued that such a machine could only perform what it was told to do; such a machine could not generate new knowledge. In 1937, Howard Aiken convinced IBM to design and build the ASCC/Mark I, the first machine of its kind, the resulting machine brought Babbage's principles of the Analytical Engine almost to full realization, while adding important new features. John Mauchly and J. Presper Eckert, two professors from University of Pennsylvania, built (1943–1944) the Electronic Numerical Integrator and Calculator (ENIAC). Considered the grandfather of digital computers, it filled a 20-foot by 40-foot room and had 18,000 vacuum tubes. The ENIAC (Electronic Numerical Integrator and Computer) was one of the first universal digital computers that was capable of being programmed. Inspired by the first computers that are as well a kind of automaton computationalism appeared. Automaton were not only a metaphor for the body but for the mind as well.

1.3.2 Computationalism and Mind

Computationalism states that the human mind is an information processing system, and that cognition and consciousness together are a form of computation. The mind is a computational system that is realized by the brain. The mind is an emergent property of complex systems simulated by a computer. The brain is viewed as being just such a complex system [205]. The computational simulation of a mind is sufficient for the actual presence of a mind, mind truly can be simulated computational on a computer. Some scientists even speculate that it is possible to achieve cybernetic immortality by downloading the information describing our brain onto a computer [205]. The computational theory of mind asserts that not only cognition, but also individual instances of subjective, conscious experience like the perceived sensation of pain, the taste of wine, as well as the redness of an evening sky are computational. Such individual instances are called qualia, derived from the Latin meaning "of what sort," such as what it is like to taste a specific apple, this particular apple now. A consequence from computationalism is the assumption that free will does not exist. Since a computer is deterministic and executes algorithms, free will does not exist since only one course of events is possible. Free will as well as the mind are illusions.

Computationalism enjoys among many researchers and philosophers a nearly orthodox status. Researchers propose that digital minds will be possible [205]. The philosopher Thomas Metzinger proposed a global moratorium strictly banning all research that directly aims at the risks suffering of artificial consciousness [192]. This is important according to him since in the feature the self-conscious machines that will exist my exceed the overall number of the humans by far.

1.4   INCOMPATIBILISM AND FREE WILL

Incompatibilism states that a deterministic universe is logically incompatible with the notion that people have free will. All physical events are caused or determined by previous events, so how can we have free will? If our deterministic brains control our behavior and produce good behaviors for us, then do we really need free choice? Is an illusion of free choice not enough? According to Dennett free will exists for us since we base our decisions on context, we limit our options as the situation becomes more specific [79]. We can as well hold persons morally responsible for their decisions because we know from historical experience that this is an effective means to make people behave in a socially acceptable way. The ethical behavior is based on the game theoretical considerations, in which moral people who cooperate will be more successful than nonmoral people who do not cooperate resulting in a beneficial mutual arrangement. Indeed, a society can define unethical standards as happened often in the human history. The human ethical heroes are the people who opted against cooperation and improvement of their society (see the Nazi Germany).

Free will is an illusion, and the mind is generated as part of the process of computation. Computationalism is based on metaphysical speculations resulting in an orthodox narrative in the scientific community. Related to this denouement is the intermixture of the terms cognition and mind. The word cognition comes from the Latin cognoscere, which means to know, to conceptualize or to recognize. Cognition is closely related to human intelligence and can be simulated by machines such as computers. On the other hand, mind has a vague definition; it has some consciousness, a combination of cognition and emotion, including an unconscious cognitive process. It manifests itself as a stream of consciousness, as described in the literary masterpiece Ulysses by James Joyce. It seems that cognition and the mind are closely related, but is this the case? To understand it better, we will study the process of computation in detail and review the principles of artificial intelligence that models human cognition. It is essential to know what computation is when dealing with the mind and brain.

CHAPTER 2

Computer Metaphor

A N algorithm is a series of steps for solving a mathematical problem, such as finding the maximum of two numbers. Such a procedure can be informal as a well, like a set of rules that describes a cookbook recipe. An algorithm is an abstract method for solving a problem described using a set of instructions, generally carried out in a sequence.

2.1 HISTORY OF ALGORITHMS

During prehistoric times and prior to the availability of written records, humans created images using cave paintings that were frequently located in areas of caves that were not easily accessible. These paintings were assumed to serve a religious or magical purpose or to represent a method of communication with other members of the group [72, 73]. Sumer is the earliest known civilization in the historical region of southern Mesopotamia, today between the Iran–Iraq border. It emerged during the early Bronze Ages between the fifth and sixth millennium BC. As human societies emerged, the development of writing was primarily driven by administrative and accounting purposes. Approximately six thousand years ago, the complexity of trade and administration in Sumer, Mesopotamia outgrew human memory. Writing became a necessity for recording transactions and administrative tasks [233,291]. The earliest writing was based on pictograms; it was subsequently replaced by letters that represented linguistic utterances [231]. More complex patterns were inscribed on larger pieces of clay. A fully formed writing system was introduced, the cuneiform script. The writing allowed the extension of the human memory as well as the communication over space and time. Approximately 4000 years ago, the "Epic of Gilgamesh," which was one of the first great works of literature, appeared. It is a Mesopotamian poem about the life of the king of Uruk [241]. The dominance of text endured in modern times until 21st century. We will follow the history of the algorithms according to the beautiful book "Poems That Solve Puzzles: The History and Science of Algorithms" by Chris Bleakley [36]. The first algorithms were written down on the clay tablets [36,38]. For example, a tablet presents an algorithm for calculating the length and width of an underground water cistern for students learning mathematics [36]. Babylonia was an ancient cultural area based in central-southern Mesopotamia. A Babylonian clay tablet around 1800 to 1600 BCE indicates the Pythagorean theorem,

DOI: 10.1201/9781003244547-2

1000 years before the ancient Greek mathematician Pythagoras was born. It seems that the Pythagorean rule was in widespread use during these times. The method of Heron of Alexandria's approximation algorithm to compute the square root of two was described (c. 10–70 CE), however the same method was already used by 1500 years before by the Babylonians [36]. The Hellenic took the leadership in mathematic based on the Mesopotamian mathematical knowledge [36]. Claudius Ptolemy (100–170 AD) was a mathematician, astronomer, geographer, astrologer and music theorist who also founded a research institute known as the Mouseion, or Museum. Its most famous building was the Library of Alexandria that represented a repository for all known knowledge represented by papyrus scrolls, a roll of papyrus containing writing. Papyrus is a material based on the papyrus plant similar to thick paper that was used in ancient times as a writing surface. It is estimated that $40,000$ to $400,000$ papyrus scrolls were housed at its height at the Library of Alexandria. Euclid (325–265 BC) was the greatest scholar at the Mouseion. Most of his writings are now lost. His great work was "The Elements" [95] about Euclidean geometry and elementary number theory. It is a mathematics textbook based on the writings of his predecessors. Euclid's algorithm is described in the "The Elements," it is an efficient method for computing the greatest common divisor (GCD) of two integers (numbers). It is the largest number that divides them both without a remainder. According to the example of [36] the divisors of 12 are 12, 6, 4, 3, 2, and 1. The divisors of 18 are 18, 9, 6, 3, 2, and 1. It follows that the greatest common divisor is 6. The Euclidean algorithm proceeds in a series of steps such that the output of each step is used as an input for the next on, it is one of the great algorithms being effective and highly efficient [36].

The original algorithm is formulated as follows according to [36]:

- "Take a pair of numbers as input.

- Subtract the smaller from the larger.

- Replace the larger with the value obtained.

- If the two numbers are equal, then output one of the numbers, it is the GCD, else apply this algorithm to the new pair of numbers."

The preceding example:

- 18, 12.

- $6 = 18 - 12$.

- 6, 12.

- $6 = 12 - 6$.

- 6, 6, GCD=6.

In the third century BCE, Eratosthenes (276–195 BCE) was appointed director of the Library of Alexandria. Eratosthenes invented an algorithm for finding prime numbers, the Sieve of Eratosthenes. There are infinitely many primes, but they are scattered across the number line without any obvious order. A prime number has no exact whole number divisors other than itself [36].

The Sieve of Eratosthenes algorithm is formulated as follows [36]:

- "List the numbers that you wish to find primes among, starting at 2.

- Repeat the following steps:

  - Find the first number that hasn't been underlined or crossed out. Underline it.

  - Cross out all multiples of this number.

  - Stop repeating when all the numbers are either underlined or crossed out.

- The underlined numbers are prim."

The number π is defined as the ratio of a circle's circumference to its diameter. Writing down the true value of π is impossible. The Babylonians had estimated $\pi = 3.125$. The Greek mathematician Archimedes of Syracuse (287–212 BC) was considered to be the greatest mathematician of ancient history, and one of the greatest of all time [36]. Archimedes' algorithm for approximating π is based on three points [36]:

- "A regular polygon approximates a circle.

- It is easy to calculate the perimeter of a polygon because its sides are straight.

- The more sides a regular polygon has, the closer its approximation to a circle."

The oldest extant Chinese mathematical book "Zhoubi Suanjing" dates to around 300 BCE focuses on the calendar and geometry, and includes the Pythagorean theorem [71].

"The Compendious Book on Calculation by Completion and Balancing" also known as Al-Jabr is an Arabic mathematical treatise on algebra written by the Muḥammad ibn Mūsā al-Khwārizmī (c.780–c.850) around 820 CE [36, 38]. Muḥammad ibn Mūsā al-Khwārizmī was a Persian polymath who produced vastly influential works in mathematics, astronomy and geography. Around 820 CE, he was appointed as the astronomer and head of the library of the House of Wisdom in Baghdad. The house of Wisdom, also known as the Grand Library of Baghdad, was a public academy and intellectual center in Baghdad during the Islamic Golden Age.

The English word "algebra" comes from the book's Arabic title (Al-Jabr meaning completion). The book provided an exhaustive account of solving for the positive roots of polynomial equations up to the second degree [38]. Kūshyār Daylami (971–1029) was a Persian mathematician, geographer and astronomer. He wrote the book "Psrinciples of Hindu Reckoning" [161] describes the decimal number system. The decimal number system itself was discovered by the Indus Valley civilization (now southern Pakistan) around 2600 BCE. The English word "algorithm" comes from the book's Latin translation "Algoritmi de Numero Indorum." Although Muḥammad ibn Mūsā al-Khwārizmī also wrote a book about Hindu arithmetic around 825 CE, his Arabic original was lost.

Performing algorithms manually is slow and leads into errors. An algorithm should be deterministic, that is why is suitable for mechanization. The first mechanical device was the abacus invented by the Sumerians around 2500 BCE. The bead-and-rail abacus was invented in China before and popularized in Europe by the ancient Romans. The first mechanical calculator (Pascaline) was invented by Blaise Pascal in 1642 as stated in the section about "Automaton and Calculators."

2.2 GÖDEL'S INCOMPLETENESS THEOREM

Can we describe mathematical proofs by an algorithm, step by step, indicating all the necessary mathematical transformations? Such an algorithm would be based on a formal system used for inferring theorems from axioms according to a set of rules that describe the corresponding mathematical transformations. Mathematics is built on axioms, considered self-evidently true but unable to be proven. The Greek mathematician Euclid (325–265 BC) wrote "The Elements" [95] about solid Euclidean geometry, elementary number theory, with a foundation of merely five axioms. Using these axioms, he proved elementary theorems about triangles, parallelograms and the Pythagorean theorem. One of these axioms is that exactly one (unique) line can be drawn through any two different points. This axiom cannot be proven, but no one doubts its veracity.

Bertrand Russell and Alfred North Whitehead completed the monumental "Principia Mathematica" [293]–a three-volume work published in 1910, 1912, 1913 and 1925–1927 which sought to provide a solid foundation for all mathematics. David Hilbert (1862–1943), a German mathematician, expressed concern that an axiomatic basis for a mathematical system might contain subtle inconsistencies leading to a paradox. The liar paradox is known from old times. For example, Aristotle mentioned it in "Sophistici Elenchi" [13]. It is based in the context of the language of self-reference. It denotes a statement that refers to itself or its own referent. The most famous example is the liar sentence: "This sentence is not true." The liar sentence leads to a contradiction when we try to determine whether it is true. It qualifies as a paradox. A paradox is a seemingly sound piece of reasoning based on apparently true assumptions that leads to a contradiction. In mathematics, one axiom might conflict with another in such a way that the conflict could manifest itself in a theorem so the theorem could be proven both true and false, a paradox. At the International Congress of Mathematicians in Paris in 1900, Hilbert presented an influential list of 23 unsolved problems in mathematics. Number two on the list was a challenge regarding the axioms: To prove that they were not contradictory, that is, that a definite number of logical steps based upon them could never lead to contradictory results. As late as 1930, Hilbert believed that there would be no such thing as an unsolvable problem.

One of the most important mathematical results of the 20th century was published by Kurt Friedrich Gödel (1906–1978), as described in the paper, "On Formally Undecidable Propositions of Principia Mathematica and Related Systems" [111], published

Figure 2.1 Kurt Friedrich Gödel (1906–1978) was considered during his time the most significant logician in history.

in 1931. Since then, Kurt Gödel has been considered the most significant logician in history (see Figure 2.1).

The results of his paper are as follows:

- Any consistent axiomatic mathematical system will contain theorems that cannot be proven.

- If all the theorems of an axiomatic system can be proven, then the system is inconsistent; thus, it has theorems that can be proven both true and false.

Mathematics cannot be represented completely by a formal system. Certain claims in mathematics are true but cannot be proven.

2.2.1 Roger Penrose's Revelation

For Roger Penrose, this paper contained a stunning revelation, as he describes in his book "The Emperor's New Mind" [212]. It told him that whatever is going on in our mind is not computational.

A mathematician makes judgments about which mathematical statements are true. If an individual can be simulated by a computer, then one could write a computer program that exactly duplicates the individual's behavior. However, any program that infers mathematical statements can infer no more than can be proven within an equivalent formal system of mathematical axioms and rules of inference. An axiomatic system cannot simply determine its own self-consistency. Penrose believes in the Platonic world of ideas where mathematics has an eternal existence. The mind is capable of contact to some extent with this eternal world. This phenomenon also explains the deep underlying reason for the accord between mathematics and physics.

2.3 TOURING MACHINE

In continuation of Hilbert's 23 unsolved problems in mathematics, David Hilbert and Wilhelm Ackermann posed a challenge in 1928, the Entscheidungsproblem. The challenge was posed before Kurt Friedrich Gödel published his ground breaking results.

The problem asks for an algorithm that considers, as input, a statement and answers "Yes" or "No" according to whether the statement is universally valid, i.e., valid in every structure satisfying the axioms.

The origin of the Entscheidungsproblem goes back to the 17th century. Gottfried Leibniz (1646–1716) constructed a mechanical calculating machine, and assumed that future machines could manipulate symbols in order to determine the truth values of mathematical statements.

It was commonly believed that there was no such thing as an unsolvable problem. Every mathematical sentence could be proved true or false.

Around 1936, Alonzo Church and Alan Turing (Figure 2.2) discovered independently that a general solution to the Entscheidungsproblem is impossible. They both showed that it is impossible to decide algorithmically whether statements in arithmetic are true or false. This result is now known as the Church-Turing Theorem.

Alonzo Church was Turing's PhD supervisor, but both followed different paths. Alonzo Church developed the λ-calculus [57–59] which was the basis for one of the first programming languages, LISP. It is a rather complicated method for defining recursive functions.

Alan Turing created a much more simpler and elegant method of mechanical computation. The method is based on a simple abstract model called the Turing machine [282]. The Touring machine became the fundament of the modern computation and algorithm analysis. The Turing machine constitutes an infinitely long tape that is divided into a sequence of cells that can represent information by some symbols and a head that can move along the tape. In each cell, a certain symbol can be written and later read by the head. A set of rules specifies in which direction the head must move depending on the symbol in a cell it reads and if it must write or read a symbol, see Figure 2.3. This set of rules specifies a new state given the current state specified by the symbols on the cells of the tape. For each state, only one rule describes one action. The computation is represented by the states of the tape. The Turing machine is a mathematical construct since the tape is infinite. However, we can simulate a restricted Turing machine with a finite long tape by a toilet paper roll and some stones representing the symbols, see Figure 2.4. One can encode the transformation rules of any specific Turing machine as some pattern of symbols on the tape that fed into a universal Turing machine. The universal Turing machine can simulate any specific Turing machine by reading in the pattern, specifying the transformation rules for that specific Turing machine. Any algorithmic process can be simulated on a universal Turing machine. In other words, we can program the Turing machine. A Turing machine represents an abstract model of a modern

Figure 2.2 Alan Turing (1912–1954) was considered eccentric by disregarding social conventions. Turing's mother could not cope with him being gay. He was not just regarded as a criminal, but as a mental case as well. It seems that in 1952 gay men were not thought sane.

computer. Anything that can be computed on a computer, can be computed on a Turing machine as well. Computationalism states that we can simulate the mind by such a universal Turing machine. The states of the mind could be represented by the states of the tape and we could simulate the mind by some stones which we move around on the toilet paper roll.

The Entscheidungsproblem corresponds to the halting problem. Whether a program will halt or run forever can be determined and is based on the given program with a finite input.

The proof is based on Cantor's diagonal argument and a form of coding in which a function or a program can be represented by a number [171]. This form of coding is called Gödelization. It allows self-reference, which means that the code of the program that is represented by a number can form an input to the program. The program can make statements about itself. The proof is based on Reductio ad absurdum, it is the form of argument that attempts to establish a claim by showing that the opposite scenario would lead to absurdity or contradiction. This technique has been used throughout history in both formal mathematical and philosophical reasoning.

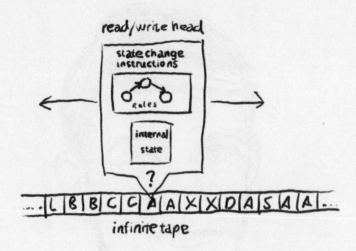

Figure 2.3 The Turing machine constitutes an infinitely long tape that is divided into a sequence of cells. In each cell, a certain symbol can be written and later read by a head. The head can move along the tape. A set of rules specifies a new state given the current state and the symbol being read. The new state determines in which direction the head must move and if it must write or read a symbol. For each state, only one rule describes one action.

Figure 2.4 A restricted Turing machine with a finite tape can be represented by a toilet paper roll and some stones that represent the symbols.

Let us suppose that a program exists that can determine if a program halts or runs forever and that this program uses a self-referent function $halt(x)$. The number x is its representation using the Gödelization principle. The function $halt(x)$ returns a one if the corresponding program represented by the number x with the input x

halts and otherwise returns a zero.

$$halt(x) = \begin{cases} 1 & program\ x\ halts\ on\ input\ x \\ 0 & otherwise \end{cases}$$

Using self-reference, we define the program with the name "Cantor"

```
Cantor(x)
{
 if halt(x)=0 then halt;
 else loop forever;
}
```

The program "Cantor" is represented by the number u. Does the program Cantor(u) halt? If yes, then we have a contradiction to the definition of $halt(x)$. If it does not halt, then there is also a contradiction to the definition of $halt(x)$. From the contradiction, it would follow that there is no program that solves the halting problem.

2.3.1 Halting Problem and the Kolmogorov Complexity

Can we use a Touring machine (a computer) to find regularities in a pattern? Let us look at a simple example of a string composed of 0 and 1,

000010000100001000010000100001000010000100010000100001

can be very shortly described using its regularities. Namely, one can say just repeat 12 times 00001. The opposite case can be seen with the string

01001110010100110110101000011101011110110110101011100100011100.

This string is basically a random sequence and since no patterns can be found no compression can be achieved. Can we find regularities in a pattern by an algorithm? The basic idea is to try to find the shortest program of the Touring machine that produces the pattern. The shortest program of the Touring machine that produces the pattern is called the Kolmogorov complexity. Because of the halting problem we cannot determine such a short program [105], no algorithmic method exists that is able to find regularities in a pattern. For example, there is no algorithm that could find regularities in a sequence of prime numbers

$$2, 3, 5, 7, 11, 13, 17, 19, 23, 29, 31, 37, 41, 43, 47, 53, 59, 61, 67, 71, 73, 79, 83, 89, 97, \cdots$$

or indicate that no such regularities exist. Despite the fact many mathematicians try to find some kind of order in the sequence of prime numbers. Another result is that a Turing machine cannot generate a truly random sequence of numbers. In some experiments, it was shown that humans as well cannot generate random sequences [249]. Randomness can be defined by numbers in a sequence [313]. A random number alone does not exist. A Turing machine is deterministic. Because of that, no true randomness exists in its context. Some facts could appear to be random. However, this arrangement is only the case because some essential information is missing.

A sequence could look random even though it is generated by a simple nonlinear deterministic equation. The Kolmogorov complexity measures of the amount of innate randomness of a sequence. The larger the shortest program that generates the sequence is, the more random it is.

2.4 COMPLEXITY THEORY

The Turing machine lead to a revolution and the beginning of computer science. Beside the question what can be computed and what not there are the questions of which problems can be solved in a finite amount of time on a Turing machine (a computer). Space and energy are negligible when using the Turing machine because the Turing machine itself is composed of infinitely long tape and does not require any energy resources [171]. To simplify the analysis of the correspondence to time, special computational problems are investigated, namely decision problems. A decision problem is a computational problem with instances formulated as a question with a binary "Yes" or "No" answer. An example is the question of whether a certain number n is a prime number. Most problems can be converted into a decision problem. The time complexity describes the amount of computer time it takes to run an algorithm. Time complexity is commonly estimated by counting the number of elementary operations performed by the algorithm. The amount of time taken is linearly related to the number of elementary operations performed by the algorithm. A problem is easy if a certain Turing machine can determine the instances related to the input for the answer "Yes" in polynomial time. Polynomial-time algorithms are said to be fast since they can be executed in an acceptable time on a computer and are called P.

Otherwise, we state that the problem is hard, means the time required grows exponentially and cannot be executed in an acceptable time on a computer, such a problem is called NP. For such a problem a Turing machine can verify in polynomial time if one instance of many represents a solution, for example, if 1023 is a solution, but no algorithm on Turing machine exits that determines the solution efficiently in time. A problem is called $NP - complete$ if all possible different computational paths must be examined by a Turing machine. A Turing machine can only solve a problem by checking all possible instances one after the other one. No other algorithm exists, we cannot speed up the computation. If we have a binary number of sixteen digits there are $2^{16} = 65,536$ possible digits which we have to test, one after another one. The expression 2^{digits} represents the exponential grow that becomes very fast intractable, cannot be computed. It was not obvious that an $NP - complete$ problem exists. Cook-Levin described the first example of an $NP - complete$ problem, the satisfiability problem. Until recently, thousands of other problems are known to be $NP - complete$, including the well-known traveling salesman and Hamiltonian cycle problem [67]. Clearly, the class $P \subseteq NP$ is known, and it follows that $NP \neq P$ or $NP = P$; however, other relationships are not known. The class $NP - complete$ is present if the problem is in NP and every other problem in NP can be reduced to the class $NP - complete$.

2.5 CHURCH-TURING THESIS AND THE CHURCH-TURING-DEUTSCH THESIS

The definition of P and NP does not depend upon the computational model. This is stated in the Church-Turing thesis: Any algorithmic process can be simulated on a Turing machine. The extended Church-Turing thesis, which is also called the strong Church-Turing thesis, states that everything that can be computed in a certain amount of time on any physical computer can be also be computed on a Turing machine with a polynomial (negligible) slowdown. In other words, any reasonable algorithmic process can be simulated on a Turing machine, with the possibility of a polynomial slowdown, in the number of steps required to run the simulation. The problems in P are precisely those for which a polynomial-time solution is the best possible, in any physically reasonable model of computation. The hypothesis that the universe is equivalent to a Turing machine, which is related to the Church-Turing thesis, is similar to that stated in digital physics. If the universe is equivalent to a Turing machine, then of course the computationalism hypothesis would be true as well. However, Richard Feynman observed in the early eighties that it did not appear possible for a Turing machine to simulate certain quantum physical processes without incurring an exponential slowdown. This fact would contradict the strong Church-Turing thesis, which led Feynman to ask whether a quantum system can be simulated on an imaginary quantum computer. The Church-Turing thesis was reformulated by David Deutsch [81] in 1985 as "Every finitely realizable physical system can be perfectly simulated by universal computing machine operating by finite means." The Turing machine was replaced by the universal computing machine which operates by finite means. Not only quantum computers are not covered by Church-Turing thesis but as well the analog computers that were popular in the 1950s. Analog computers cannot be simulated by a Turing machine.

2.5.1 Analog and Digital Computers

An analog computer represents information by analog means, such as voltage. In such a computer, information is represented by a voltage wave and the algorithm is represented by an electrical circuit. Such a circuit is composed of resistors and capacitors that are connected together. An algorithm represents a mathematical model of a physical system, which can be described, for example, by specific differential equations. The input and output of the computation are voltage waves that can be observed by an oscilloscope. The represented values are usually less accurate than digitally represented values. The results of each computation can vary due to external influences. For this reason, each result of the computation is unique. The exact value cannot be reproduced without an error. This type of noise, which results from an external influence, makes it impossible to recompute and reproduce exactly the output for certain functions. Even making the smallest change to the initial condition can cause the results to greatly diverge. Analog computers correspond to a computing machine that operates by finite means and are only covered by the Church-Turing-Deutsch thesis. Analog computers fell into decline with the advent of the development of the microprocessor, which led to the development of digital computers and a possible reproduction of a calculation without an error.

A digital computer is a device that processes information that is represented in discrete means such as symbols. Usually, the symbols are represented in binary form. Modern digital computers are based on digital circuits and the von Neumann architecture. In a digital circuit, the information is represented by binary digits. Due to a digital representation, the exact values of each computation can be reproduced since there is no external influences. The computation can be repeated, and the result remains the same. Binary digits are represented by the minimal unit of information, the bit with the values 0 or 1. The binary information is manipulated by Boolean digital circuits that are composed into the on the von Neumann architecture build up of five main modules, see Figure 2.5:

- Input and output mechanisms that communicate with the user.

- A control unit (CU) that interprets an instruction retrieved from the memory and that selects alternative courses of action based on the results of the previous operations.

- Arithmetic and logic operations on the data are performed by an arithmetic logic unit (ALU)

- Main memory stores both data and instructions and read-write random-access memory (RAM).

- Secondary memory represents the external mass storage.

The von Neumann architecture is based on the Electronic Numerical Integrator and Calculator (ENIAC) build and developed (1943–1944) by John Mauchly and J. Presper Eckert. They proposed the EDVAC's (Electronic Discrete Variable Automatic Computer) construction in August 1944. Unlike its predecessor the ENIAC, it was binary rather than decimal, and was designed to be a stored-program computer. John von Neumann learned in 1945 of the ENIAC and EDVAC project. He described later EDVAC model in a technical report called "First Draft of a Report on the ED-VAC" [286]. The failure of von Neumann to list John Mauchly and J. Presper Eckert, as authors on the First Draft led credit to be attributed to von Neumann alone, since then the model become known as the Van Neumann architecture [16].

2.5.1.1 Digital Computers

In the beginning, digital computers were designed for companies and for large research centre. Now there are included in ordinary mobile phones. The digital camera of a mobile phone has nothing to do with the film camera. It takes remarkable computing power in a phone to take photos or make videos. The most important part are in the algorithms, for example, they analyze the photograph to determine what is in front and what is behind. Todays networks were are associated with computers, allowing them to communicate with each other. The search engine is such a staggering invention that no one remembers how it was done before. Object are connected together

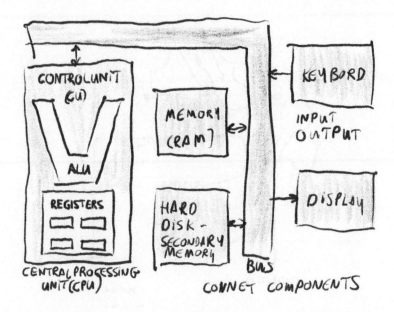

Figure 2.5 In Van Neumann architecture, the ALU and the CU are parts of the central processing unit (CPU) represented by a single silicon chip called a microprocessor. The subsystem called a bus transfers the data between the random-access memory, the CPU and the input and output.

by computer networks. A car has essentially become an algorithms on wheels. The brakes and the motor are controlled by algorithms. Science is more and more based on simulations. To simulate is to build on a computer a kind of computer model of the process that we want to study.

2.6 ARTIFICIAL INTELLIGENCE

Cybernetics has its origins in the intersection of the fields of control systems, electrical network theory, mechanical engineering, logic modeling, fuzzy logic, evolutionary biology, neuroscience, anthropology and psychology in the 1940s, often attributed to the Macy Conferences. In the early 1940s, John von Neumann added the von Neumann cellular automata and the von Neumann Universal Constructor with the concept of self replication.

In 1943, McCulloch, a neuroscientist, and Walter Pitts, a logician, developed the artificial neuron, a mathematical model that mimics the functionality of a biological neuron. This model is called the McCulloch-Pitts model of a neuron. The nervous signals enter the neuron through dendrites and exit through its single axon. The dendrites receive signals from other neurons, the soma processes the information and the axon transmits the output. The synapses are the points of connection between neurons where memory traces are stored, see Figure 2.6. In the spring of 1947a congress on harmonic analysis held in Nancy, Norbert Wiener began to write a manuscript on applied mathematics and telecommunication engineering. The book was published 1948 with the title "Cybernetics: Or Control and Communication in the Animal and

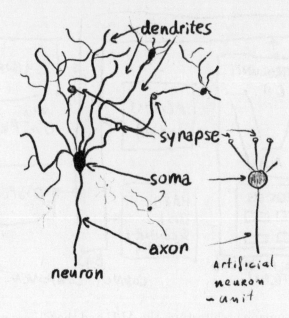

Figure 2.6 A simple representation of a biological neuron. The dendrite receives signals from other neurons, the soma processes the information, the axon transmits the output of this neuron and the synapse is the point of connection to the other neurons. On the right, we see an abstract model of a biological neuron, a unit or an artificial neuron.

the Machine" [311]. It was the first public usage of the term "cybernetics" to refer to self-regulating mechanisms and provided a foundation for research in computing.

Alan M. Turing (1912–1954), wrote in 1950 the essay "Computing Machinery and Intelligence" [281], in which he poses the question of how to determine whether a program is intelligent or not [281]. He defines intelligence as the reaction of an intelligent being to certain questions. This behavior can be tested by the so-called Turing test. A subject communicates over a computer terminal with two non-visible partners, a program and a human. If the subject cannot differentiate between the human and the program, the program is called intelligent. For Turing, a machine thinks if its behavior is indistinguishable from human intelligence. Turing was also involved in the birth of computational chess, by introducing the Turochamp. Turochamp was a paper machine, a set of rules indicating a program of how to play chess using paper and pencil. It is the earliest known computer chess game to enter development, but was never completed by Turing, as its algorithm was too complex to be run on the early computers.

The perceptron algorithm was invented in 1957 by Frank Rosenblatt [232] and was inspired by the McCulloch-Pitts model of a neuron. Perceptron describes an algorithm for supervised learning that considers only linearly separable problems in which groups can be separated by a line or hyperplane, see Figure 2.7. In the 1960s, an active research program concerning machine learning with artificial neural networks

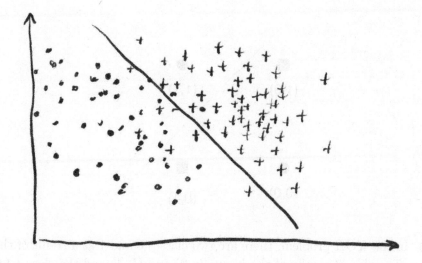

Figure 2.7 In a sample, each object of class A is represented by a dot and each object of class B by a cross. We separate the two classes using a straight line. Some errors are present, but most examples are correctly separated [89]. Later, during classification, an unknown class is described by a two-dimensional vector, and depending on which side of the separating line it falls, it is classified as either a class A or class B.

was carried out by Rosenblatt. After a press conference with Rosenblatt, the New York Times published an article in 1958 in which they claimed: "The Navy revealed the embryo of an electronic computer today that it expects will be able to walk, talk, see, write, reproduce itself and be conscious of its existence.... Later, perceptrons will be able to recognize people and call out their names and instantly translate speech in one language to speech and writing in another language."

Artificial intelligence (AI) was founded as a distinct discipline at the Dartmouth workshop in 1956. The term itself was invented by the American computer scientist John McCarthy and used in the title of the conference. During this meeting, programs were presented that played chess and checkers, proved theorems and interpreted texts. Arthur Samuel developed machine learning algorithms for checkers. Checkers requires intelligence when the algorithm for playing is unknown. As soon as the algorithm is known, playing checkers no longer requires intelligence. Artificial intelligence (AI) is a subfield of computer science that models the mechanisms of intelligent human behavior (intelligence). This approach is accomplished via simulation with the help of artificial artifacts, typically with computer programs on a machine that performs calculations.

After some uneasy coexistence with cybernetics, AI gained funding and prominence. Cybernetic sciences such as the study of artificial neural networks like the perceptron were downplayed.

Rosenblatt's schoolmate at the Bronx High School of Science Marvin Minsky and his colleague, Seymour Aubrey Papert, published a book 1969 with the title

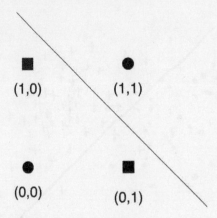

Figure 2.8 In the XOR problem, there are two classes 0 and 1 as a result of the XOR operation. The class 0 results of the input (0, 0) and (1, 1) and the class 1 form the input (0, 1) and (1, 0). The classes cannot be separated by a straight line.

"Perceptrons" [194]. In the book, it was argued that perceptrons cannot model non-linear functions such as the logical XOR operation (see Figure 2.8). Shortly after the book was published, Rosenblatt died in a boating accident. Some people assume it was suicide; he was definitely heartbroken. As a consequence of the published book, nearly no research was conducted in artificial neural networks for the next 10 years. Instead research in symbolic artificial intelligence was carried out.

In the science fiction film from 1968, "2001: A Space Odyssey" by In Stanley Kubrick's (1928–1999), a computer was introduced with the name HAL (abbreviation for heuristic algorithms) that illustrated the goals of AI. HAL understood spoken language, conducted dialogues, played chess, and solved planning problems [312]. HAL, the imaginary computer of 1968 could easily pass the Turing test [264]. The Limited Turing Test was passed on May 11, 1997 through a chess-playing program Deep Blue. The Deep Blue program developed by IBM beat the world chess champion Garry Kasparow 3.5–2.5 in six games [199]. In 2011, "Watson" a question-answering computer system capable of answering questions posed in natural language developed by IBM competed on Jeopardy! against champions Brad Rutter and Ken Jennings, winning the first place prize of 1 million dollar [98]. AlphaGo a game playing algorithm (Go/Chess) that is based on reinforcement learning [68]. In the Future of Go Summit held in Wuzhen in May 2017, AlphaGo Master won all three games with Ke Jie, the world No.1 ranked player.

2.7 SYMBOLICAL ARTIFICIAL INTELLIGENCE AND THE THEORY OF MIND

The key idea behind symbolic artificial intelligence is the symbolic representation of the domain in which the problems are solved. Symbols are used to denote or refer to something other than themselves, namely other things in the world according to the,

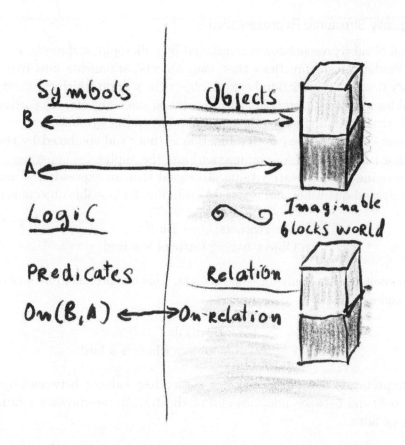

Figure 2.9 Objects are represented by symbols and the relations by predicates. In our figure, the predicate $ON(A, B)$ represents the relation between two blocks represented by the symbols A and B.

pioneering work of Tarski [270–272]. In this context, symbols do not, by themselves, represent any utilizable knowledge. For example, they cannot be used for a definition of similarity criteria between themselves [202, 256] (see Figure 2.9).

The computational theory of mind requires that mental states are representations. The use of symbols in algorithms which imitate human intelligent behavior led to the famous physical symbol system hypothesis by Newell and Simon (1976) [200]: "The necessary and sufficient condition for a physical system to exhibit intelligence is that it be a physical symbol system." Symbols are not present in the world; they are the constructs of a human mind and simplify the process of representation used in communication and problem solving.

In Fodor's original views, the computational theory of mind is also related to the language of thought, published 1975. The language of thought hypothesis allows the mind to process more complex representations with the help of semantics, thinking occurs in a mental language. One way of such representation is the logically structured representation.

2.7.1 Logically Structured Representation

Logical symbolical representation is motivated by philosophy and mathematics [160, 176, 272]. Predicates are functions that map objects' arguments into true or false values. They describe the relation between objects in a world which is represented by symbols. Whenever a relation holds with respect to some objects, the corresponding predicate is true when applied to the corresponding object symbols.

Predicates can be negated by the function $\neg$ (not) and combined by the logical connectives $\vee$ (disjunction), $\wedge$ (conjunction) and the implies ($\rightarrow$) operator. $\neg$, $\vee$, $\wedge$, and $\rightarrow$ determine the predicate's value. To signal that an expression is universally true, the universal quantifier and a variable standing for possible objects is used.

$$\forall x[\text{Feathers}(x) \rightarrow \text{Bird}(x)].$$
An Object having feathers is a bird.

Some expressions are true only for some objects. This is represented by an existential quantifier and a variable.

$$\exists x[\text{Bird}(x)].$$
There is at least one object which is a bird.

An interpretation is an accounting of the correspondence between objects and object symbols and between relations and predicates. An interpretation can be only either true or false.

2.7.2 Rules

A rule [176, 236, 315] contains several "if" patterns and one or more "then" patterns. A pattern in the context of rules is an individual predicate which can be negated together with arguments. The rule can establish a new assertion by the "then" part, the conclusion whenever the "if" part, the premise, is true. When variables become identified with values they are bound to these values. Whenever the variables in a pattern are replaced by values, the pattern is said to be instantiationed. Here is an example of rules with a variable x:

- If $\underbrace{(\text{flies}(x) \vee \text{feathers}(x)) \wedge \text{lays eggs}(x)}_{premise}$ then $\underbrace{\text{bird}(x)}_{conclusion}$

- If bird(x) $\wedge$ swims(x) then penguin(x)

- If bird(x) $\wedge$ sings(x) then nightingale(x)

The following assertions are present:

- feathers(Woodstock)

- lays eggs(Woodstock)

- sings(Woodstock)

- flies(Red Baron)

- barks(Snoopy)

Woodstock is a bird because the premise of the first rule is true when x is bound to Woodstock. Because bird(Woodstock), the premise of the third rule is true and Woodstock is a nightingale.

2.7.3 Search in a Problem Space

The problems in AI are often described by the representation of a problem space and a search procedure [204]. The abstract representation of a problem by the search for valid moves in the state space is one of the most important methods of AI. The Swiss mathematician Leonhard Euler (1707–1783) introduced the abstract representation of the relationships of the world in 1735 by investigating the problem of the connections of the bridges in the city of Königsberg. The city is on two sides of a river with two islands in the middle. The islands and both banks of the river are connected by seven bridges. To investigate whether a stroll through the city is possible in which each bridge is only crossed once, Euler introduced the graph theory [66]. It is an abstract representation of the problem by the state space in which the current position of the walker is represented together with the transitions between the bridges. To answer the question of whether such a walk is possible, a search for all possible routes is carried out.

From 1958 to 1969, Newell and Simon worked on the General Problem Solver (GPS) [94, 255]. A problem is described by a set of states and a set of rules which convert the states into other states. A rule can be executed if its specified preconditions can be met. A problem is specified by an initial state and goal states. In addition, a difference table is given, which can be used to calculate the difference between the current state and the target state. The rules are executed whose preconditions are met and which minimize the difference to the target states. This heuristic technique is called the means-ends-analysis (MEA). Heuristics are strategies that can accelerate the search for a solution. The GPS program uses a representation of a problem space as well as a heuristic search. Production systems like GPS are an important tool in AI [256]. They consist of rules (corresponds to long-term memory) and a working memory in which the description of the world is located. The aim is to achieve a target state based on a starting state. This is done with the help of a search strategy. Production systems are used to model the knowledge of experts in the so-called expert systems.

2.7.3.1 Expert Systems

Expert systems simulate the conclusions of a human expert when solving problems with the help of specialist knowledge. In contrast to general approaches of artificial intelligence, expert systems work on the basis of knowledge; they differ from other knowledge-based systems in that the knowledge comes from an expert [104, 141, 151, 175, 183]. The motivation for the expert systems comes from observing the nature of human intelligence when reasoning on a narrowly limited field of activity. Human expertise depends on the field in which it is created. Human experts know more, but they don't think faster or in any other way than non-experts. Their performance results from success in finding solutions to these tasks. One of the most important features of the architectures of expert systems is the separation between the knowledge the associated inference methods. The knowledge that human experts use must be extracted and formulated through discussions and relevant specialist books [2]. This task is carried out by knowledge engineers. The extracted and formulated knowledge provides a detailed model of the knowledge base of the human experts. A human expert is someone who has acquired specific skills and knowledge in a particular area, like a doctor [296], lawyer [240], mineralogist, paleontologist [298], psychoanalyst.

We demonstrate an example of the expert system "Jurassic" [298]. The goal of the system is to help paleontologists determine creatures or creature groups out of a taxonomic knowledge base which describes the dinosaurs [122] and using only some vague beliefs about the presence and absence of some features. In 1887, Professor Harry Govier Seeley grouped all dinosaurs into the saurischia and ornithischia groups according to their hip design. The saurischian were divided later into two subgroups: The carnivorous, bipedal theropods and the plant eating, mostly quadruped sauropodomorphs. The ornithischians were divided into the subgroups birdlike ornithopods, armored thyreophorans and margginoncephalia. The subgroups can be divided into suborders and then into families and finally into genus. The genus includes the species. It must be noted that in this taxonomy many relations are only guesswork, and many paleontologists have different ideas about how the taxonomy should look. The whole knowledge base is composed of 70 modules in which 423 rules are stored. The taxonomy is modeled in the books of [163, 164] in which over fifty families and more than 340 genre of dinosaurs are described. Which object of one group is most similar to an object of another different group? We wish to know which animal of the dinosaur group is most similar to an animal of the mammal group. This is a kind of case-based reasoning [128] in which specific knowledge is used to retrieve the most similar stored case. "Which dinosaur species is most similar to a human?" We describe a human being by seven features. The categorization is performed in which the most similar stored case is determined.

1. Certainly **two legged**.

2. Certainly not **four legged**.

3. Probably **thin walled fragile bones**.

4. Very probably not **big**.

5. Very probably **big eyes**.

6. Certainly **big brain**.

7. Certainly **capsule in the skull**.

8. Very probably **length two m**.

R E S U L T : **STENONYCHOSAURUS**

This answer is the same as the suggestion of Dale Russell [234] in the early 1980s that the Stenonychosaurus (also now known as Troodon), could have given rise to a brainy descendant, had dinosaurs survived instead of dying out [163, 164]

If these experts use their skills in public, it looks so effortless that one is tempted to attribute special skills and talents to them. However, when psychologists began to measure the general intelligence of the experts with the help of psychological tests, they did not find any general superiority over the non-experts. The superiority of the experts often relates to a very limited area, the transfer of the expertise to other areas is limited or hardly possible. It is generally believed that experts have more knowledge of a subject than non-experts. The performance of the human experts increases through experience, whereby the acquired knowledge plays a central role. However, the assumption of increased performance through the acquisition of knowledge from experience often does not apply. Sometimes beginners are just as good as or even better than experts if they have the appropriate knowledge, e.g. in the form of instructions. Beginners who start with their experience in a field change their behavior and increase their performance for a period of time until they reach an accepted level. Beyond this state point, however, it is not possible to make a prediction as to whether an increase in performance will occur or not. The human capacity to absorb and organize knowledge appears to be limited. Above all, experts extract the most important information from memory. A difference between a beginner and an expert is not only in the amount of knowledge acquired, but also in the way that knowledge is organized. The knowledge of an expert is organized around central terms of a field in order to guarantee quick access to relevant information. In the case of less experienced people, the knowledge is arranged around everyday terms, which makes it difficult to find certain knowledge content. These observations led to the assumption that the procedure of human experts is mechanical in nature, that is, that corresponding processes can be automated with the help of computer programs. For this purpose, the knowledge that the human expert needs is to be extracted and represented by a program and the associated inference methods are simulated.

The software architecture of the expert systems is based on the separation of knowledge and reasoning mechanisms. The knowledge is represented in a so-called knowledge base, the inference mechanisms are implemented by an inference engine. This separation allows the knowledge to be presented in a natural form, making it easier to maintain the knowledge base. The inference engine determines what is to be done in a particular situation, depending on the current state. There are different ways to represent the knowledge declaratively, like predicate logic (logic) or rules. Rules are the most common form of the representation of knowledge. These rules are closer to the representation used by human experts, e.g. rules in the form "If X then Y." The rule interpreter of the inference engine decides in which order these rules are used during problem solving. Problems without side effects of actions can be described by deduction systems which are a subgroup of production systems [315]. In deduction systems the premise specifies combinations of assertions, by which a new assertion of the conclusion is directly deduced. This new assertion is added to the memory. Deduction systems may chain together rules in a forward direction, from assertions to conclusions, or backward from hypotheses to premises. During backward chaining, it is ensured that all features are properly focused. Backward chaining is used if no features are present. If all features are given, forward chaining is used to prevent wasting of time pursuing hypotheses which are not specified by the features. The chained rules describe the complete problem space which can be represented by a semantic net [225, 251]. For clarity, rules can be arranged in groups [2, 145] which define a taxonomy according to their dependence between the conclusion pattern and one premise pattern.

The software architecture is implemented by a so-called expert system shell. This consists of various modules which, on the one hand, enable the knowledge base to be created and maintained by the knowledge engineer and, on the other hand, enable dialogues with the user during problem solving. The knowledge component is used to create and check the consistency of the knowledge base as well as its modification and maintenance. The problem solving component interprets the knowledge which is necessary for the solution of problems specified by the user. The interview component conducts dialogues with the user or reads in automatically collected data. The explanation component makes the procedure of the expert system transparent, i.e. that the expert system can explain and justify its behavior to a certain extent. The user has the possibility to ask the questions how (HOW) and why (WHY) to the system during the dialogue. To HOW, the expert system answers how it has reached a certain state, to the question WHY, why it asked a certain question to the user [175]. The expert systems are used in categorization of knowledge, configuration, design, monitoring of systems (e.g. nuclear power plants), troubleshooting, prognosis, diagnosis as well as teaching and training [215, 240]. A major challenge of expert systems emerges when the size of the knowledge base increases due to contradictory knowledge and knowledge acquisition [162]. Today expert systems are mainly used for diagnostic problem solving in industrial machines [299, 300] or in relation in on board diagnostics (OBD) in cars. Expert systems indicate us that the knowledge is independent of the interference mechanism, the process of reasoning is independent of the knowledge, it and can be described by knowledge independent algorithms.

2.7.4 Sub-Symbolic Representation and Intuition

The sub-symbolical or distributed representation became popular with the books Parallel Distributed Processing (PDP) published by David Rumelhart and McClelland [186–188]. Sub-symbolical representation tries to overcome the problems of symbolical representation by the heuristic functions or intuition resulting from distributed representation that measures the similarity of the represented objects. It implies and explains the heuristic functions or intuition since we can measure similarity between different sub-symbols. Mental imagery problem solving is an example of human intelligent behavior, which is not performed by a physical symbol system but rather by a sub-symbolic manipulating system, which mimics the sensory representation of the world [159]. The sub-symbolical representation often corresponds to a pattern that mirrors the way the biological sense organs describe the world. Patterns are represented by vectors, which can be classified according to categories. Categories themselves can be represented by symbols. In this sense, the vectors correspond to sub-symbols, which are transformed into symbols through the elimination of information [301]. A vector is only a sub-symbol if there is a relationship between the vector and the corresponding similarity of the represented object or state in the real world through sensors or biological senses.

2.7.5 Cognitive Models of Human Thinking

It is quite certain for the cognitive psychologist that the human computational model is not at all based on the van Neumann architecture. Instead, they propose that the human computational model is based on production systems. A production system is a mathematical as well as a practical model that can be realized as a computing machine. Production systems are equivalent in power to a Turing machine [282]. A Turing machine can also be easily simulated by a system. The production system is a model of actual human problem solving behavior [9, 151, 201, 202].

Production systems are composed of if-then rules that are also called productions. A production contains several "if" patterns and one or more "then" patterns. A pattern in the context of productions is an individual predicate, which can be negated together with arguments. A production can establish a new assertion by the "then" part (its conclusion) whenever the "if" part (its premise) is true. A production system is composed of [45, 176]:

- The human long-term memory is modeled by a set of productions.

- The human short-term memory or working memory that represents the mental states. This memory contains a description of the state in a problem solving process. The state is described by logically structured representation and is simply called a pattern. Whenever a premise is true, the conclusions of the productions change the contents of the working memory. The mental state is changed.

- The focus of attention, also called the recognize-act cycle. If several productions can bye applied to the working memory the focus of attention determines which production to use. This choice can be based on a kind of intuition also called a heuristic function. Such an heuristic function can be learned from experience, if in similar problems a production leads to success. A heuristic function can as well estimate the cost of reaching a goal by using a certain production.

The computation is done in the following steps. The working memory is initialized with the initial state description. The patterns in working memory are matched against the premise of the production. The premise of the productions that match the patterns in working memory produces a set, which is called the conflict set. One of the productions of this set is chosen using the focus of attention and the conclusion of the production changes the content of the working memory. This process is denoted as firing of the production. This cycle is repeated on the modified working memory until a goal state is reached or no productions can be fired. The production system in the context of classical Artificial Intelligence and Cognitive Psychology is one of the most successful computer models of human problem solving. The production system theory describes how to form a sequence of actions, which lead to a goal, and offers a computational theory of how humans solve problems [7]. One can as well build production systems that are based on sub-symbolic representation.

2.7.5.1 Examples

The 8-puzzle is composed of eight numbered movable tiles in a 3×3 frame. One cell of the frame is empty; as a result, tiles can be moved around to form different patterns. The goal is to find a series of moves of tiles into the blank space that changes the board from the initial configuration to a goal configuration.

The long-term memory is specified by four productions [176]:

- **If** the empty cell is not on the top edge, **then** move the empty cell up;

- **If** the empty cell is not on the left edge, **then** move the empty cell left;

- **If** the empty cell is not on the right edge, **then** move the empty cell right;

- **If** the empty cell is not on the bottom edge, **then** move the empty cell down.

The control strategy for the search would be:

- Halt when goal is in the working memory.

- Chose a random production.

- Do not allow loops.

In Figure 2.10, we see an example representing a sequence of states that lead form the initial configuration to the goal configuration.

Another example consists of the task of building a tower from a collection of blocks [204]. A robot arm can stack, unstack, and move the blocks within a plane on three different positions at a table. There are two different classes of blocks: Cubes and pyramids. While additional blocks may be stacked on top of a cube, no other blocks may be placed on top of a pyramid. The robot arm, which is represented in the upper right corner, has a gripper that can grasp any available block. It can move the block to eight different positions on the tabletop or place it on top of another cube [294, 297, 310]. In Figure 2.11, we see a planing with a random choice of the production called the blind search. With the aid of this heuristic, heuristic search is performed. Out of several possible productions, we chose the one, which modifies the visual represented state in such a way that it becomes more similar to the visual represented goal state according to the Euclidean distance [301], see Figure 2.12. During the blind search, an impasse can be present where there is no valid transition to a succeeding state. An impasse is solved by backtracking to a previous state, see Figures 2.13 and 2.14.

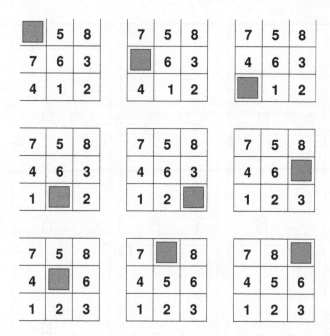

Figure 2.10 The first pattern (upper left) represents the initial configuration and the last (low right) the goal configuration. The series of moves describes the solution to the problem.

2.7.5.2 SOAR

One of the best-known cognitive models, based on the production system, is SOAR. The SOAR state, operator and result model was developed to explain human problem-solving behavior [202]. SOAR is a theory which attempts to unify all the theories of the mind in a single framework. It is a unified theory of cognition that proposes mechanisms by which the results of these human cognitive experiments can be reproduced. SOAR describes the architecture of the mind: A fixed structure underlying the flexible domain of cognitive processing that models the human behavior. All problem solving activity is formulated as the selection and application of productions to a state, to achieve some goal. The decision takes place in the context of earlier decisions. Those decisions are rated utilizing preferences and added by chosen rules. Preferences are determined together with the rules by an observer using knowledge about a problem. SOAR models the psychological phenomena of chunking, the association of expressions or symbols (chunks) into a new single expression or symbol (chunk). Chunking represents a theory of learning. A chunk is a new rule (production) that describes the processing that was present due to lack of applicable knowledge. A lack of applicable knowledge is present if it cannot be decided which rule to use or no rule can be applied for a certain mental state represented in the working memory. SOAR models as well the perceptual and motor subsystem that consist of independent modules for each output channel. They run asynchronously with respect to each other. The perceptual modules deliver data directly into working memory whenever it is available. The motor modules accept commands from working memory and execute them. The SOAR

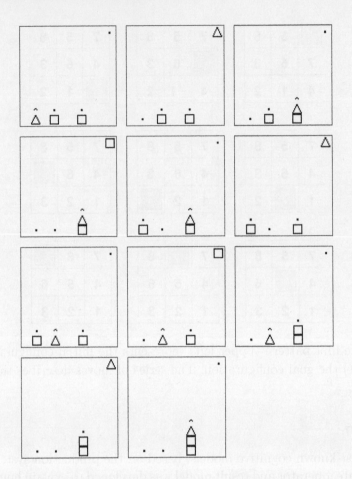

Figure 2.11 The simplest method corresponds to a random choice of a production. An example of planning of the tower building task of three blocks. The upper left pattern represents the initial state; the bottom right pattern, the goal state.

architecture can be and is used for intelligent agents like, for example, autonomous robots (Robo-Soar). SOAR simulated pilots flying U.S. tactical air missions in simulation, it was also used to simulate virtual humans supporting face-to-face dialogues and collaboration within a virtual world.

2.8 COMPLEX SYSTEMS

A complex system is a system composed of many parts which interact with each other. An example of a complex system is the human brain where many neurons interact which each other [90,247]. Complex systems can be simulated on a computer; however they are difficult to model due to their relationships, such as nonlinearity interactions between them and their environment that may result in feedback loops. A complex system is defined as a set of entities that, through their interactions, relationships, or dependencies, form a unified whole resulting in behaviors which are distinct from

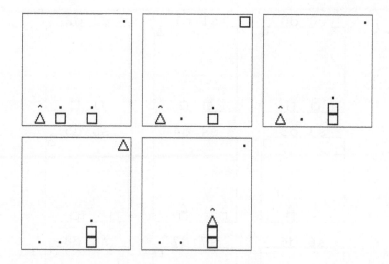

Figure 2.12 An example of planning of the tower building task of three blocks using the heuristic search. The upper left pattern represents the initial state; the bottom right pattern, the goal state.

the properties and behaviors of its parts. A system exhibits complexity when the systems' behaviors cannot be easily inferred from its properties.

2.8.1 Self-Organization

We are often surprised when observing natural phenomena like smoke-rising by its regular behavior [156]. The regularity comes from self-organization, a process where some form of overall order arises from local interactions between parts of an initially disordered system. Self-organization of patterns and structures can be described by synergetics.

2.8.1.1 Synergetics

The synergetic means in Greek cooperation. Synergetics itself is an interdisciplinary field that deals with cooperation of the individual parts of the system that may lead to formation spatial or temporal structures [123, 124]. For example, when fluid is heated more strongly it may form patterns. The heat is transported on a microscopic (small scale) level. When the temperature difference between the lower and upper surfaces of the liquid exceeds a critical value, macroscopic (large scale) motion can be observed in form of rolls. The fluid rises at specific positions, then it cools down at the upper surface and sinks down at different position in a circular way and an ordered pattern emerges. see Figure 2.15.

The system forms a new structure by self-organization. The formation of the structure can be interpreted as the emergent properties of the system. The macroscopic order is independent of the details of the microscopic interactions of the subsystems.

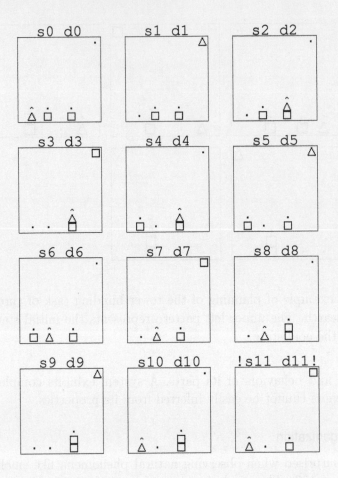

Figure 2.13 An impasse is present when there is no valid transition to a succeeding states. The upper left pattern represents the initial state, above the image *s* indicates the steps and *d* the depth of the search. An exclamation point after *d* indicates that an impasse is present. In Figure 2.14 the search continues.

2.8.1.2 Associative Memory

A complex system that is based on self-organization is the associative memory that is modeled by interacted neurons.

The associative memory incorporated the following abilities in a natural way [7, 133, 156, 206]:

- The ability to correct faults if false information is given.

- The ability to complete information if some parts are missing.

- The ability to interpolate information. In other words, if a sub-symbol is not currently stored the most similar stored sub-symbol is determined.

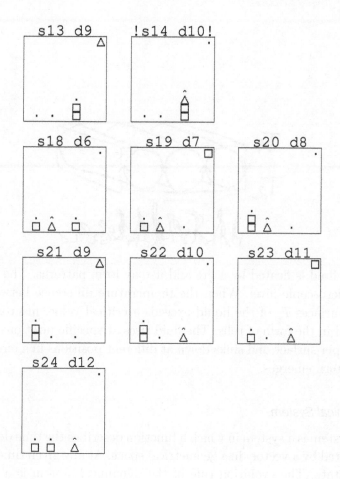

Figure 2.14 Continuation of the search represented by Figure 2.14. The bottom right pattern represents the goal state.

When an incomplete pattern is given, the associative memory is able to complete it by a dynamical process.

"Human memory is based on associations with the memories it contains. Just a snatch of well-known tune is enough to bring the whole thing back to mind. A forgotten joke is suddenly completely remembered when the next-door neighbor starts to tell it again. This type of memory has previously been termed content-addressable, which means that one small part of the particular memory is linked - associated -with the rest." Cited from [46], page 104.

The Hopfield network represents a model of the associative memory [133, 136]. In a Hopfield network, all units are connected to each other by the weights, see Figure 2.16. Patterns that the network uses for training (called retrieval states) become attractors of the system. After the training, we start with some configuration and the network will converge after several steps using the update rule to an attractor representing a fixed point (a vector), see Figure 2.17.

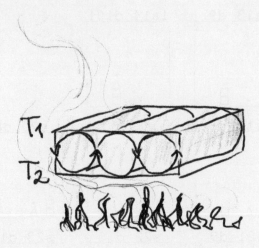

Figure 2.15 The fluid is heated by a fire and it may form patterns. The heat is transported on a microscopic level. When the temperature difference between the lower T_2 and upper surfaces T_1 of the liquid exceeds a critical value, macroscopic motion can be observed in the form of rolls. The fluid rises at specific positions, then it cools down at the upper surface and sinks down at different positions in a circular way and an ordered pattern emerges.

2.8.1.3 Dynamical System

A dynamical system is a system in which a function describes the time dependence of a point (represented by a vector) in a geometrical space. At any given time, a dynamical system has a state. The evolution rule of the dynamical system is a function that describes what future states follow from the current state. An attractor is a state or a set of states toward which a system tends to evolve from a starting condition of the system [20]. A Hopfield network is a dynamical system. After the training we start with some configuration represented by a binary vector (a point) and the network will converge after several steps using the update rule to an attractor representing the stored pattern or a spurious state composed of a linear combination of an odd number of stored patterns. An attractor can be a fixed point or a cycle. However, there are other attractors that cannot be easily described as simple combinations that are called strange attractors. An attractor is called strange if it has a fractal structure. Fractals exhibit similar patterns at different scales, this property is called self-similarity. Strange attractors are related to the chaos theory.

2.8.1.4 Chaos Theory

Chaos theory describes underlying patterns and deterministic laws highly sensitive to initial conditions in dynamical systems that appear to have a random evolution. Chaos theory states that within the apparent randomness there are underlying patterns. The butterfly effect, a principle of chaos theory, describes how a small change in one state of a deterministic nonlinear system can result in large differences in a later state [109]. A metaphor for this behavior is that a butterfly flapping its wings

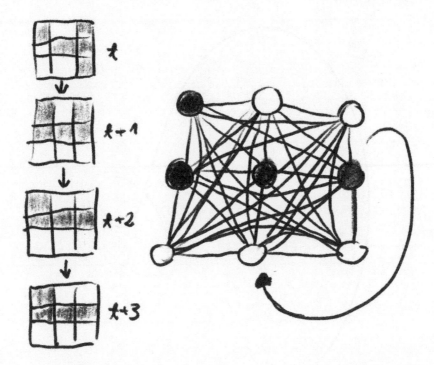

Figure 2.16 In a Hopfield network all units are connected to each other by weights. Each unit has a value -1 indicated by a white pixel or 1 indicated by a black pixel. All the activation of the units represent a pattern at state t. Usually the units are updated asynchronously, updated one at a time t. After the training we start with some configuration and the network will converge after several steps using the update rule to an attractor if the state at t is the same at $t + 1$.

Figure 2.17 Example of how the Hopfield network restores a distorted pattern. The network will converge after several steps using the update rule to an attractor representing the stored pattern or a spurious state composed of a linear combination of an odd number of stored patterns.

can cause a tornado. In the absence of perfect knowledge of the initial conditions even the minuscule disturbance, our ability to predict its future course will always fail. Physical systems can be completely deterministic and yet still be inherently unpredictable even in the absence of quantum effects. The theory was summarized by Edward Lorenz a mathematician and meteorologist as: "*Chaos: When the present determines the future, but the approximate present does not approximately determine the future.*" Edward Lorenz discovered the Lorenz system is a system of three

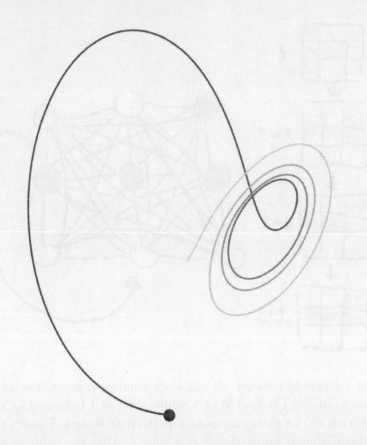

Figure 2.18 The Lorenz attractor is a set of chaotic solutions of the Lorenz system that depends on the initial condition. In the plot we show the solution for $t = 6$.

ordinary differential equations [153]. It is notable for having chaotic solutions for certain parameter values and initial conditions with the Lorenz attractor representing a set of chaotic solutions of the Lorenz system. The shape of the Lorenz attractor itself, when plotted graphically, may also be seen to resemble a butterfly, see Figures 2.18, 2.19 and 2.20.

2.8.1.5 Mandelbrot Set

The Mandelbrot set has its origin in complex dynamics. The Mandelbrot set is the set of complex numbers c for which the function $f_c(z) = z^2 + c$ does not diverge to infinity when iterated from $z = 0$ for which the sequence $f_c(0), f_c(f_c(0)), f_c(f_z(f_c(0)))), \cdots$ remains bounded in absolute value.

A complex number is an element that contains the real numbers and an imaginary unit i that satisfies the equation $i^2 = -1$. Every complex number can be expressed in the form $a + bi$, where a and b are real numbers and be represented as a vector (point) in two-dimensional space, axis representing the real and imaginary part.

On 1 March 1980, at IBM's Thomas J. Watson Research Center, Benoit Mandelbrot first saw a visualization of the set. Benoit Mandelbrot (1924–2010) was a

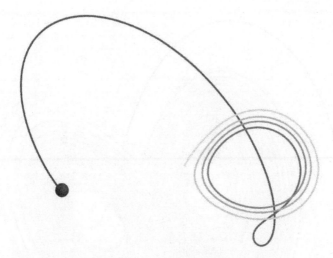

Figure 2.19 The Lorenz attractor is a set of chaotic solutions of the Lorenz system that depends on the initial condition. In the plot we show the solution for $t = 6$ with a little bit of a different initial condition than in 2.18.

Polish-born French American mathematician [180, 181]. The images of the Mandelbrot set exhibit an elaborate and infinitely complicated fractal boundary. It reveals progressively ever finer recursive detail at increasing magnifications, see Figure 2.21.

Kolmogorov complexity of Mandelbrot set is low, despite the fact a complex aesthetic structure arises by the application of simple rules.

2.8.2 Artificial Life

One example of a complex system whose emergent properties have been studied extensively is the cellular automata. In a cellular automaton, a grid of cells, each having one of the finitely many states, evolves according to a simple set of rules [318]. The rules are defined locally, despite the fact they have shown globally interesting behavior.

The British mathematician John Conway created the most famous cellular automaton the Game of Life [106, 318]. The Game of Life marks the beginning of artificial life. It is played on a square lattice. Each lattice site has a value 0 or 1. A one indicates that the site is alive, a zero that it is dead. The system evolves by updating all the sites in the lattice until two successive lattice configurations are identical, or until a specified number of updates have occurred. The rules, known as life and death rules, are based on the value of a site and the sum of values of its neighbors.

- A living site with two living nearest neighbor sites remains alive.

- Any site with three living nearest neighbor sites remains alive or is born.

- All other sites remain dead or die.

The game is played on a square lattice, with initial random set-up of zeros and ones. Then the rules are applied for each site of the square. One should note that the

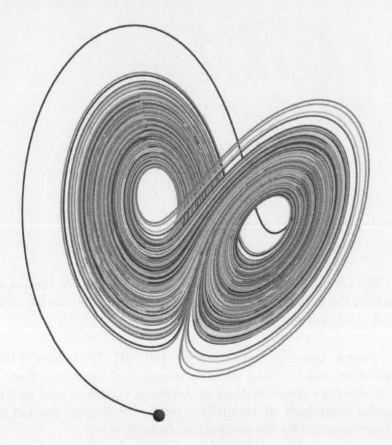

Figure 2.20 The Lorenz attractor with the same initial condition as Figure 2.18 with the solution for $t = 400$. The shape of the Lorenz attractor itself, when plotted graphically, may also be seen to resemble a butterfly.

second rule is a more specific than the first one. The system evolves from a random configuration until two successive lattice configurations are identical, see Figure 2.22. The Game of Life is of great interest when persistent patterns that move appear, known as life forms, see Figure 2.23.

Game of life is a universal cellular automaton, in the sense that it is effectively capable of emulating a Turing machine [31, 318].

Artificial life was named in 1986 by Christopher Langton, an American theoretical biologist. Artificial life examines systems related to life, its processes, and its evolution, through the use of simulations with computer models, robotics and biochemistry. It tries to recreate aspects of biological phenomena of living systems in artificial environments in order to gain a deeper understanding of the complex information processing that define such systems.

The strong artificial life states that life is a process which can be abstracted and simulated on a computer, life is not simulated but synthesized. The weak artificial life position tries to simulate life processes to understand the underlying mechanics of biological phenomena.

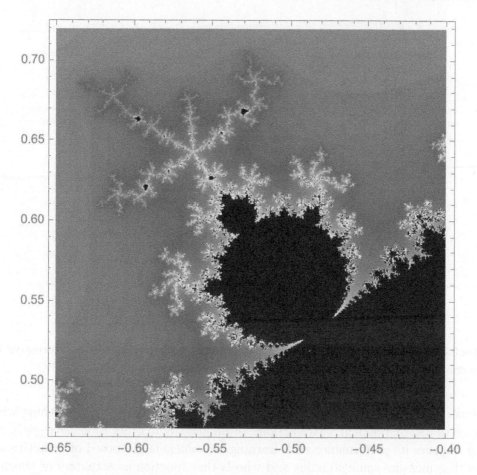

Figure 2.21 The Mandelbrot set is the set of complex numbers c for which the function $f_c(z) = z^2 + c$ does not diverge to infinity when iterated from $z = 0$. The images of the Mandelbrot set exhibit an elaborate and infinitely complicated fractal boundary. It reveals progressively ever finer recursive detail at increasing magnifications. The figure part of the Mandelbrot set is shown with the real axis -0.65 to -0.4 and the imaginary axis 0.47 to 0.72.

2.8.2.1 Genetic Algorithm

In computer science, a genetic algorithm is a heuristic search inspired by the process of natural selection simulating the evolution. Genetic algorithms used to generate solutions to optimization and search problems by relying on biologically inspired operators such as mutation, crossover and selection [100].

2.8.2.2 Braitenberg Vehicle

A Braitenberg vehicle is a concept that was introduced by the Italian-Austrian cyberneticist Valentino Braitenberg (1926–2011). A Breitenberg vehicle models animal world in a minimalistic way by simple reactive behavior resulting in the formation of concepts, spatial behavior and generation of ideas [42]. It is an agent that can

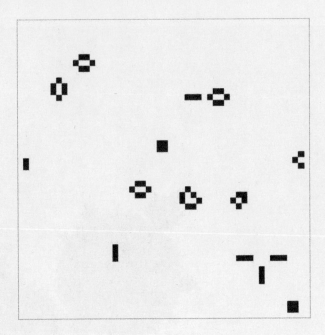

Figure 2.22 The system evolves from a random configuration until two successive lattice configurations are identical.

autonomously move around based on its sensor inputs. An agent is anything which perceives its environment, takes actions autonomously in order to achieve goals, and may improve its performance with learning. A vehicle is composed of primitive sensors that measure some stimulus and wheels that function as actuators or effectors. Each wheel has its own motor, see Figure 2.24.

Depending on how sensors and wheels are connected, the vehicle exhibits different behaviors. It appears to strive to achieve certain situations and to avoid others, see Figure 2.25.

2.8.2.3 Swarm Intelligence

A multi-agent system composed of multiple interacting agents that can solve problems that are difficult or impossible for an individual agent. Swarm intelligence is the collective behavior of decentralized, self-organized systems composed of agents [100]. The inspiration often comes from nature, especially biological systems. Examples of swarm intelligence in natural systems include ant colonies, bee colonies, bird flocking, animal herding and fish schooling.

The agents follow simple rules, and there is no centralized control structure. Through interactions between agents' intelligent global behavior emerges, unknown to the individual agents.

2.8.3 Emergence

A common feature of complex systems is the presence of emergent behaviors. Emergence occurs when an entity is observed to have properties its parts do not have

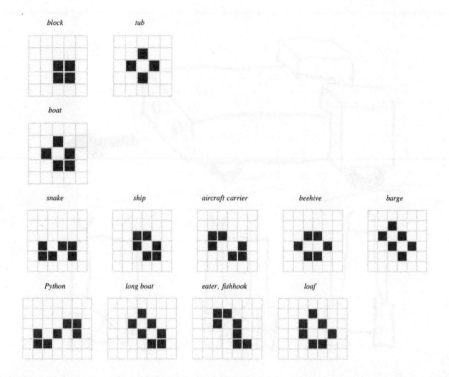

Figure 2.23 The Game of Life is of great interest when persistent patterns that move appear which are known as life forms. Some known life forms are block, tub, boat, snake, ship, aircraft carrier, beehive, barge, python, long boat, eater, fishhook and loaf.

on their own, properties or behaviors which emerge only when the parts interact in a wider whole. Emergent properties arise from the collaborative functioning of a system, but do not belong to any one part of that system.

An ant colony is composed of individual ants. Each ant is without mind and intelligence, however the behavior of the colony may seem to be intelligent and we could assume that it has its mind of its own. As a collective, ants are efficient and surprisingly intelligent. Emergent properties are the changes that occur in ant behavior when individual ants work together [100].

The actions of a single ant are almost random. But many actions by many of ants are organized and solve complex tasks like building hills and dams to finding and moving huge amounts of food.

2.8.4 Society of Mind

Marvin Lee Minsky (1927–2016) proposed the society of mind theory in which a vast society of individually simple processes known as agents interact whit each other. An agent is without mind and can be represented by a simple computational process [193]. These agents are mindless entities from which minds are built. Mind is a kind of state resulting by the phase transition from mindless simple entities to an abstract concept

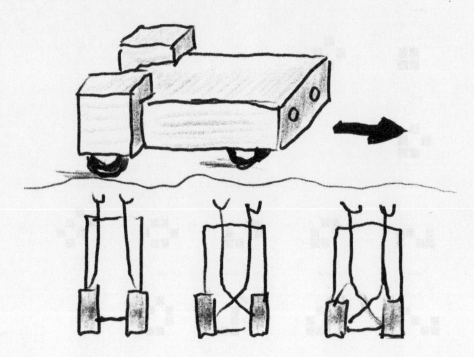

Figure 2.24 A Braitenberg vehicle is composed of primitive sensors that measure some stimulus and wheels that function as actuators or effectors. Each wheel has its own motor; the lower three patterns indicate different connections between the sensors and the motors of the two wheels.

of mind. As Marvin Minsky notice in his book "The Society of Mind" [193]: *What magical trick makes us intelligent? The trick is that there is no trick. The power of intelligence stems from our vast diversity, not from any single, perfect principle.* Society of mind can explain the appearance of intelligence, but its name intermixture intelligence with mind and consciousness. The description is related to the physical phase transitions (or phase changes) are the physical processes of transition between a state of a medium, identified by some parameters, like changes among the basic states of matter: Solid, liquid, and gas. It is also related to a hierarchical representation of abstract concepts. For example, a ship is an entity that is guided by the decisions of its captain. An alien outside observer could assume that the ship makes its own decisions by its own mind. The same would be true for the captain, he can be recursively decomposed into simple agents that compose his mind.

2.8.5 Strange Loop

Douglas R. Hofstadter that argues our self and mind is based on self-representation and an abstract feedback loop [135]. Feedback occurs when outputs are routed back as inputs, such feedback can be described by an endless recursion. It is argued that such a feedback leads to the perception of the self by its representation. The abstract self representation is required to model the causality of our actions.

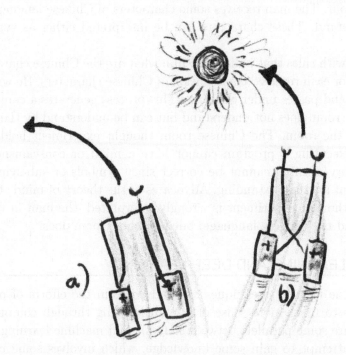

Figure 2.25 Depending on how sensors and wheels are connected, the vehicle exhibits different behaviors. It appears to strive to achieve certain situations and to avoid others. Vehicle (a) moves away from the source (light); vehicle (b) moves to the source.

2.8.6 Emergentism

Emergentism is the belief in emergence, particularly as it involves consciousness and the philosophy of mind. Consciousness is believed to appear in certain large neural networks, but is not an attribute of a single neuron. It is an emergent property of the human brain. Emergentism is compatible with the theory that the universe is composed exclusively of physical entities, and with the evidence relating changes in the brain with changes in mental functions.

2.9   CRITIQUE OF THE THEORY OF MIND

Emergent behaviors of complex system through self organization creates new patterns that can be predicted by carful mathematical analysis of the complex system. It is new and unsuspected behavior for the observer since she/he often lack the corresponding mathematical tools. Consciousness does not belong in this category since it is not a pattern. In "The Conscious Mind" [55] David Chalmers argues that consciousness is irreducible to lower-order physical facts.

Let us imagine a thought experiment known as the "Chinese room" that was proposed by John Searle [224, 245]. A lonely man in a room can only communicating with the outside world through a piece of paper with symbols written on it that is passed under the door. The man receives some characters of Chinese language which he does not understand. These characters can be interpreted either as symbols or sub-symbols.

He uses a book with rules that indicate to him what are the Chinese characters he has to write down for each received combination of Chinese characters. He writes this combination down and passes under the door. This process generates a conversation that the man in the room does not understand but can be understand by the Chinese speaker outside of the room. The Chinese room thought experiment holds that a digital computer executing a program cannot have a mind or consciousness. The computational theory of mind cannot be correct since symbols or sub-symbols are manipulated without any understanding. Advocates of the theory of mind state that the Chinese room thought experiment is wrongly interpreted, the man in the room does not understand the Chinese language, but the whole room does.

2.10 MACHINE LEARNING AND DEEP LEARNING

Many of the machine learning techniques are derived from the efforts of psychologists and biologists to make more sense of human learning through computational models [7]. There are some parallels between humans and machine learning. During learning, humans attempt to gain some knowledge, which involves some modification of behavioral tendencies by experience. In machine learning, we can distinguish between supervised learning and unsupervised learning. In supervised learning, the algorithm is presented with examples of inputs and their desired outputs. The goal is to learn a general rule that maps inputs to outputs. Supervised learning is frequently referred to as learning with a teacher because the desired outputs are indicated by a type of teacher. Consequently, unsupervised learning is referred to as learning without a teacher. In unsupervised learning, the algorithm groups information that is primarily represented by vectors into groups. The algorithm attempts to find the hidden structure of unlabeled data; clustering is an example of such an algorithm [309].

The limitations of a simple perceptron do not apply to feedforward networks with nonlinear units, also called multilayer perceptrons. Such networks can be trained by the backpropagation learning algorithm. The algorithm itself was invented independently several times, [47, 187, 210, 292]. However, it became popular with the books Parallel Distributed Processing (PDP) published by David Rumelhart and his colleagues, [186–188].

The popularization of the backpropagation algorithm through the PDP books led to a connectionist movement that resulted in the "symbol wars" with the old symbolic artificial intelligence school. The symbol wars describe the emotional discussion of the two camps around the question as to whether the departure from the symbolic approach leads to something new and worthwhile.

According to the universality theorem, a neural network with a single nonlinear hidden layer trained by backpropagation is capable of approximating any continuous function. Attempting to build a network with only one layer to approximate complex functions often requires a very large number of nodes ("fat" neural networks). The immediate solution to this is to build networks with more hidden layers. Empirical

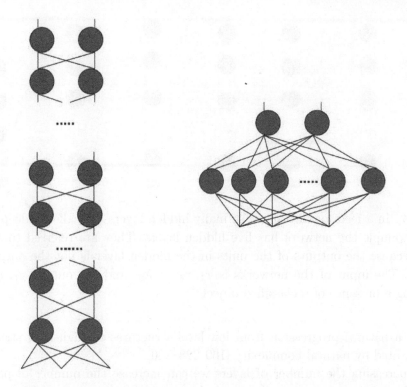

Figure 2.26 Empirical experiments indicate that "deep" neural networks (left) give better results than "fat" neural networks (right).

experiments indicate that "deep" neural networks give better results than "fat" neural networks, see Figure 2.26. The term "Deep Learning" was introduced to the machine learning community by Rina Dechter in 1986 [77] and to artificial neural networks by Igor Aizenberg and colleagues in 2000 [4].

Even the early machine learning algorithms were motivated by biology this not the case with modern machine learning algorithms like support vector machines or deep learning. They are based on statistical methods and the relation to the brain becomes less important [309]. They represent engineering tools rather than models of the human cognition. For example, it is assumed that in the brain the mass action of thousands to millions of neurons spread over the brain represents a classified object, not the activity of single output neuron [319] that indicates a certain object like in deep learning, see Figure 2.27. The presence of each classified object is indicated by an active neuron in deep learning. The output neurons are as well called the grandmother neurons, a neuron that represents a specific object, like for example your "Grandmother." If this neuron dies (drinking too much alcohol), you will be not able to recognize your grandmother.

Deep artificial network uses many hidden layers to increase the model's power in statistical learning, see Figure 2.27. These layers are trained by backpropagation. As a result after the training during the classification no explanation can be given why a certain input was mapped to a certain class. Deep learning enables high-level abstractions in data by architectures composed of multiple nonlinear transformations.

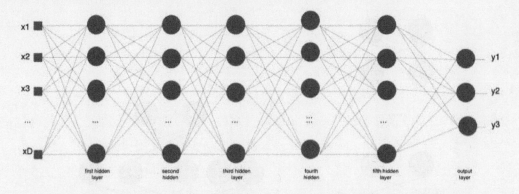

Figure 2.27 In a feed forward network many hidden layers are called a deep network. In our example the network has five hidden layers. They are referred to as hidden layers because the outputs of the units in the hidden layer is not the output of the network. The input of the networks is $x_1, x_2, \cdots X_D$ and the output y_1, y_2 and y_3 indicating a presence of a classified object.

It offers a natural progression from low-level structures to high-level structure, as demonstrated by natural complexity [169, 228–230].

By increasing the number of layers we can increase the number of parameters faster. By doing so, we can add enough degrees of freedom to model large training sets. This is extremely helpful since nowadays a really large amount of data is collected for specific tasks.

The models achieved tremendous results, and it is now common to identify artificial intelligence with deep learning and not with symbol manipulating systems. This results from the paradox of artificial intelligence in which the terms "intelligence" and "intelligent human behavior" are not very well defined and understood.

However, it seems that the deep learning revolution results mainly from brute force, it is not based on new mathematical models and appears to be biologically unlikely [248]. Deep neural networks require a very large labeled training set. This requirement can become a bottleneck, since in many special applications it is difficult to generate labels for enough examples. Labels indicate the class of each input pattern during training through backpropagation. A deep learning neural network is a black box system without the possibility of explanation or interpretation. It is also a kind of black art, there are many approaches and hyper-parameters and often these need tweaking without any mathematical explanation beside that certain configuration of these parameter give good results.

Human brains can learn from much less examples, in notoriously difficult tasks, while providing deeper understanding a generalization capability. This led the machine learning pioneer Geoffrey Hinton to suggest that entirely new methods will probably have to be invented. He stated[1]: "I don't think it's how the brain works.... We clearly don't need all the labeled data." Perhaps some of these new methods may

---

[1]Interview with Steve LeVine, Sep 15, 2017, AXIOS, https://www.axios.com.

come from a deeper understanding of biological networks that would overcome the problems of deep learning.

2.10.1 Computationalism and Machine Learning

The computational theory of mind is related to the representational theory of mind in that they both require that mental states are representations. Symbolical and sub-symbolical representation model the world that can be represented and manipulated in the short-term memory. Deep learning algorithms are black box algorithms without any meaningful representation and correspond to unconsciousness thinking. They map sensory input into symbols or sub-symbols that represent the mental state.

2.11 COMPUTER AND THE TIME PERCEPTION

Time perception of time is for us humans a subjective conscious experience like sensation of the taste of wine. It is related to qualia.

In "The Magic Mountain" by Thomas Mann (1875–1955) the main character Hans Castorp has reached the conclusion that, for the mind, time does not flow uniformly; the mind only assumes it does so to maintain the proper order of things. Therefore, all measurements of time are no more than conventions.

As Saint Augustine (354–430) quoted: "What then is time? If no one asks me, I know what it is. If I wish to explain it to him who asks, I do not know." Neither time past nor future, but present only, really is for us humans.

Computer does not represent time in the way humans do. In a computer each computation step is represented by a state. There is no difference between printing the states of a computation and a real time computation.

A Turing machine makes no explicit mention of the time scale over which computation occurs. One could physically implement the same Turing machine with a silicon-based device that preforms the computations in milliseconds, or a slow mechanical device that preforms the same computation over years or represent the computation by a printed book outside any time frame. In fact, it is difficult to explain time by computationalism.

A computer does not know that it computes since it does not perceive time. A computation can be represented by a printed book, but a book without any reader is meaningless. Only for the reader is able to experience time through the process of reading.

CHAPTER 3

Brain

T HE BRAIN is the most complex structure in the universe that we know. Can we explain its resulting psychological phenomena in terms of neurobiological phenomena? Can we describe neurobiological phenomena by algorithms? What would be the nature of such algorithms? Are there some phenomena that cannot be described by algorithms? We give a short overview from different perspectives and try to answer the questions by using some ideas that sketch the working principles of the brain.

3.1 NEURONS

The cell is the basic functional unit of all known organisms. It is the smallest unit of life and constitutes its building blocks [276]. Bacteria are free-living simple cells. They were the first forms of life to appear on Earth, approximately 4 billion years ago. Bacteria reproduce by cell division; they can move and communicate. Initial approaches have been made of simulating bacteria on a computer; however, the simulations are quite complex and push the limits of existing computers.

In complex multicellular organisms such as humans, cells specialize into different types that are adapted to particular functions. Major cell types include skin, muscle, blood and stem cells as well as neurons and fibroblasts. A neuron or nerve cell is an electrically excitable cell that processes information by communicating with other neurons. The human brain is estimated to include approximately 86 billion of these neurons–a massive number.

A nerve cell or a biological neuron consists of a cell body (soma), dendrites, axons and synapses [60]. Dendrites receive signals from other neurons, somas process the information, and axons transmit the output. The synapse connects to other neurons; it permits a neuron (or nerve cell) to pass an electrical or chemical signal to another neuron or to the target effector cell. Synapses represent memory traces since they can store information through certain changes. These simplified computational operations can be described by a simple mathematical model. Incoming signals from other neurons are represented by vector $\mathbf{x}$ and the weight of each synapse is represented by vector $\mathbf{w}$. Vector $\mathbf{w}$ represents the memory of the neuron. Both vectors have the same dimension (the same number of entries). The similarity between the two vectors is measured by the scalar product between the two vectors. If the scalar value is bigger than a certain threshold value, then input $\mathbf{x}$ is similar to stored memory $\mathbf{w}$ and the

DOI: 10.1201/9781003244547-3

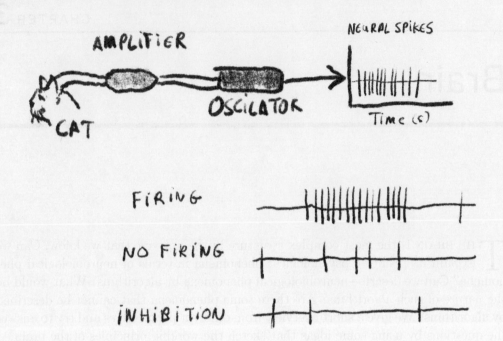

Figure 3.1 We measure the activity of a neuron, amplify it and represent it by an oscillator. A firing neuron corresponds to spikes with high frequency, a non-firing neuron releases some uncorrected spikes from time to time. Also, there are inhibitory neurons that correspond to negative information; in this case there are no spikes in a certain period [137].

neuron fires. Firing indicates that the neuron sends output information to the other neurons to which it is connected. Otherwise, the neuron does not fire, and no information is sent. Sending information and not sending information can be represented by a one and a zero, respectively. Such simple models are used in machine learning.

In reality, the operation is more complex since a neuron firing corresponds to several spikes at a high frequency, and a neuron not firing, by spikes at a low frequency. In Figure 3.1, the analysis of electrical signals is recorded by electrodes set in the brain of a cat. We measure the activity of a neuron, amplify the information and represent it by an oscillator [137]. A firing neuron correspond to spikes with high frequency, a non firing neuron releases from time to time some uncorrected spikes. There are also inhibitory neuron that correspond to negative information, in this case there are no spikes in certain period. A continuous activation can be represented by different frequencies. Every firing signal is referred as a spike, or an action potential [92]. An action potential occurs when a neuron sends information down an axon, away from the cell body. The action potential is an explosion of electrical activity that is created by a depolarizing current, see Figure 3.2. When the threshold level is reached, an action potential of a fixed sized will always fire. Information is stored in connections between cells, the synapses. When two neurons are active at the same time, the connection (synapse) between them is strengthen. The learning

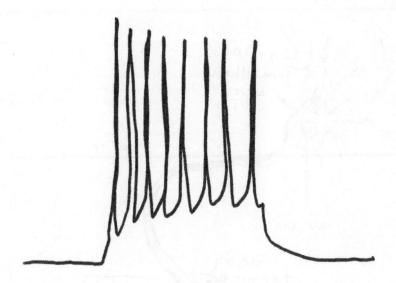

Figure 3.2 An action potential occurs when a neuron sends information down an axon, away from the cell body. The action potential is an explosion of electrical activity that is created by a depolarizing current. The x-axes correspond roughly to 50 ms.

rule is called the Hebbian learning rule [130, 131]. The Hebbian learning rule specifies how much the weight of the connection between two units should be increased or decreased. The Hebbian learning is the basis of some machine learning algorithms like the Willshaw's associative memory [314]. It is a local rule since only the information of neighboring neurons is used to adapt the weights of the synapses. Local learning rules are biologically plausible, contrary to backpropagation algorithm used in deep learning.

There are two types of synapses, the electrical and the chemical [276]. In an electrical synapse, the presynaptic and postsynaptic cell membranes are connected by special channels called gap junctions and are capable of passing an electric current, causing voltage changes in the presynaptic cell that induce voltage changes in the postsynaptic cell. The main advantage of an electrical synapse is the rapid transfer of signals from one cell to the next. Chemical synapses allow neurons to form circuits within the cerebral cortex and are crucial to the biological computations that underlie perception and thought. In Figure 3.3, we represent a neuron with a cell body, the soma, dendrites and an axon. The soma contains the nucleus that controls and regulates the activities of the neuron. The axon transmits the output information of the neuron via the chemical synapse by releasing neurotransmitter molecules into a small space, called the synaptic cleft, adjacent to another neurons. In a chemical synapse, electrical activity in the presynaptic neuron is converted into the release of a chemical called a neurotransmitter. Neurotransmitters are chemical messengers that transmit a signal from a neuron across the synapse to another neuron. Neurotransmitters are released from vesicles in synapses into the synaptic cleft, where they are received by neurotransmitter receptors on the target neuron. They are transported over the

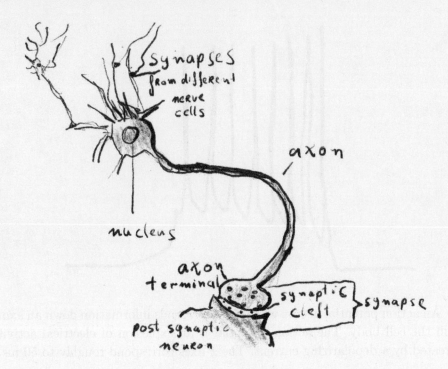

Figure 3.3 A neuron with a cell body, the soma, dendrites and an axon. The soma contains the nucleus that controls and regulates the activities of the neuron. The axon transmits the output information of the neuron via the chemical synapse by releasing neurotransmitter molecules into a small space, called the synaptic cleft adjacent to other neurons.

synaptic cleft by reuptake transporters, a class of proteins that span the cellular membranes of neurons. The autoreceptor is located in the membranes of presynaptic nerve cells and serves as part of a negative feedback loop inhibiting further release or synthesis of the neurotransmitter. A released neurotransmitter is typically available in the synaptic cleft for a short time before it is absorbed by enzymes, see Figure 3.4. Neurotransmitters are essential to the function of complex neural systems. More than 500 unique neurotransmitters in humans were identified till today. The chemical synapse is a complicated mechanism to determine the weight (memory) of the synapse trough some chemical context information.

3.1.1 Mind and Drugs

The behavioral neuropharmacology focuses on the study of how drugs affect the human mind. Drugs can reinforce neurochemical reward system. Synapses are affected by drugs such as curare, cocaine, morphine, nicotine, alcohol, LSD and others. Certain drugs are restricted to synapses that use a specific neurotransmitter by altering its activity [276]. Morphine acts on synapses that use endorphin neurotransmitters, and alcohol increases the inhibitory effects of the neurotransmitter (GABA). LSD interferes with synapses that use the neurotransmitter serotonin. Cocaine blocks reuptake

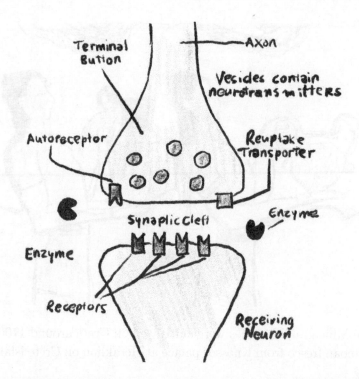

Figure 3.4 Neurotransmitters are released from vesicles in synapses into the synaptic cleft, where they are received by neurotransmitter receptors on the target neuron. They are transported over the synaptic cleft by reuptake transporters and they are a class of proteins that span the cellular membranes of neurons. The autoreceptor is located in the membranes of presynaptic nerve cells and serves as part of a negative feedback loop inhibiting further release or synthesis of the neurotransmitter. A released neurotransmitter is typically available in the synaptic cleft for a short time before it is absorbed by enzymes.

of dopamine and therefore increases its effects. Drugs targeting the neurotransmitter of major systems affect the whole system, which can explain the complexity of action of some drugs like cocaine. Cocaine causes pleasurable feelings by elevating dopamine concentrations in the synapses of the reward system. With repeated exposure, some glutamate receptors in the reward system become sensitized to cocaine cues. Repeated drug use leads to changes in neuronal structure and function that cause long-lasting or permanent neurotransmission abnormalities. Morphine counteracts pain and triggers great joy and high spirits. It is precisely this euphoric effect that seems to be responsible for the fact that morphine and in particular its derivative heroin have become severely abused drugs. Buspirone sold under the brand name Buspar treats symptoms of anxiety and depression. Buspirone was first made in 1968 and approved for medical use in the United States in 1986. On the other hand, Reserpine is a drug that is used for the treatment of high blood pressure, but it leads as well to sedation and depression.

Figure 3.5 The Minoan noble wines were cultivated on Crete around 1700 BC. This is an ancient Minoan fresco from Knossos palace at Heraklion on Crete island in Greece.

3.1.1.1 Drugs as Stimulants

Some psychoactive drugs have traditionally been viewed as stimulants in parts of the world and are consumed by broad sections of society. These include alcohol, like beer or wine, nicotine like tabac and caffeine [276]. Alcohol usage results in sedation, memory impairment and muscle relaxation. As early as 5000 BC the wine appeared for the first time in the South Caucasus (today Georgia) [189] and in the Middle East Sumer (today southern Iraq). The Minoan noble wines were cultivated on Crete around 1700 BC [127], see Figure 3.5.

Nicotine like tobacco smoking increases the attention and concentration. The tobacco plant originally comes from America and the consumption was already known long before the European conquerors set foot on the continent [56]. In 1500, the Portuguese navigator Pedro Cabral encountered pipe smoking and in 1518 a very developed smoking culture in Mexico was discovered by the Spaniards. Caffeine, like coffee, increases wakefulness. Coffee was mention already as early as the 9th century in southwest Ethiopia, Africa.

3.2 DIVISION OF NERVOUS SYSTEM

There are different levels of investigation of the nervous system that are related to the spatial scale, from the molecular level to the entire central nervous system. At each level different structures and functions are found, from molecules to synapses and neurons itself to networks, maps and systems [97, 184]. Higher level description by computer simulations often neglect the lower levels. Neurons are abstracted and the learning in neural networks is described by simplified mathematical methods. Higher

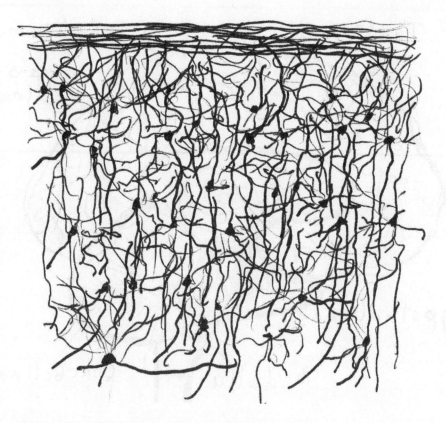

Figure 3.6 Neurons from the visual cortex of a rat.

level systems are often described by symbolical rules. This is done due to extensive complexity that is beyond our present computational models [276].

The human nervous system can be divided into the central nervous system that corresponds to the brain and the spinal cord. The human behavior is not only described by the brain, but as well by the spinal cord and the sensory neurons. They implement several reflexes that lead into automatic behavior, like the withdrawal reflex when we touch something hot and withdraw our hand from the source. You are cooking and touch a hot plate, the hand is withdrawn at once, much later you feel the pain. Indeed, our mind can override such reflexes in certain situations.

The brain is of the central interest when dealing with the question of mind. The neurons in the brain form networks that preform individual tasks, see Figure 3.6. It is estimated that the human brain accounts for around 86 billion of these neurons. There are many different types of neurons. To understand the function of such networks, we analyze the brain functions like indicated in Figure 3.7. The limbic system supports a variety of functions including emotion, behavior, long-term memory and smell (not shown in Figure 3.7). It is a set of brain structures located on both sides of the thalamus.

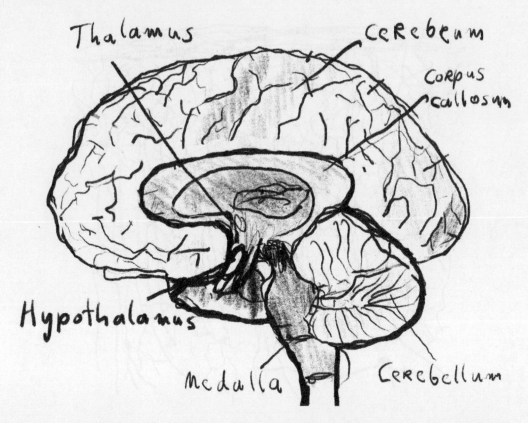

Figure 3.7 The medulla controls heart rate, breathing, peristalsis and reflexes such as sneezing. The hypothalamus controls temperature and water homeostasis, the internal water state that should be steady maintained. The hypothalamus also controls the release of hormones by the pituitary gland. The thalamus is a relay station integrating sensory input and channelling it to the sensory areas of the cerebrum. The cerebrum is divided into two hemispheres that are connected by the corpus callosum (see Figure 3.8). The cerebellum coordinates muscle movement and thus controls balance, posture and movement. It is thought that complex motion is composed of simpler movements called elemental movements. The cerebellum learns to execute complex movements by associating patterns of context information with elemental movement patterns.

3.2.1 Endocrine System

In addition to the nervous system, the endocrine system is the second largest communication system of the human body where the information is passed through blood that is circulated through the vertebrate vascular system. While the nervous system uses neurotransmitters as its chemical signals, the endocrine system uses hormones [276]. The pancreas, kidneys, heart, adrenal glands, gonads, thyroid, parathyroid, thymus and even fat are all sources of hormones. The endocrine system works in large part by acting on neurons in the brain, which controls the pituitary gland. The hormones activate specific receptors in their target organs, which then release other hormones into the blood, which in turn affect other tissues and the brain. The

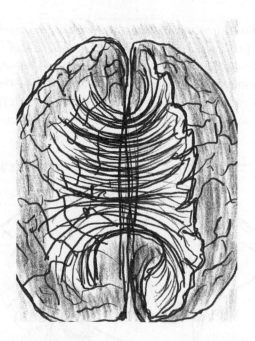

Figure 3.8 The cerebrum is divided into two hemispheres that are connected by the corpus callosum.

endocrine system is both responsible for activating and controlling basic behavioral functions such as sexuality, emotions and stress functions and for regulating body functions such as growth. In response to stress and night cycles, hormones enter the blood and travel to the brain and other organs and the brains capacity for neurotransmission are changed over a course of hours to days. By doing so, the brain adjusts its performance and control of behavior in response to a changing environment.

3.2.2 The Cerebrum

In the human brain, the cerebrum is the uppermost region of the central nervous system and the neocortex is the largest part of the cerebral cortex. Neocortex is composed of the outer layer of the cerebrum and is made up of six layers, labelled from the outermost inwards. The cerebrum contains the cerebral cortex of the two cerebral hemispheres and several subcortical structures controlling emotions, hearing, vision, personality and much more. The two hemispheres remain in contact and communication with one another by the corpus callosum, see Figure 3.8. Functionally, left cerebral hemisphere controls right side of body and right cerebral hemisphere controls left side of body. Perceptual information is processed in both hemispheres, information from each side of the body is sent to the opposite hemisphere. The two hemispheres are roughly mirror images of each other, with only subtle differences. Some of these hemispheric distribution differences are consistent across human beings, many observable distribution differences vary from individual to individual, like the hand preference (left or right hand) that is related to hemisphere specialization (lateralization). The lateralization of brain function is the tendency for some neural functions or cognitive processes to be specialized to one side of the brain or the other.

The cerebrum is divided into many subdivisions and sub-regions as indicated in the Figure 3.9. The primary sensory areas of the cerebral cortex receive impulses via sensory neurons from receptors that detect the stimuli reaching the body. The skin has more receptors in some parts of the body than others. Motor areas send impulses

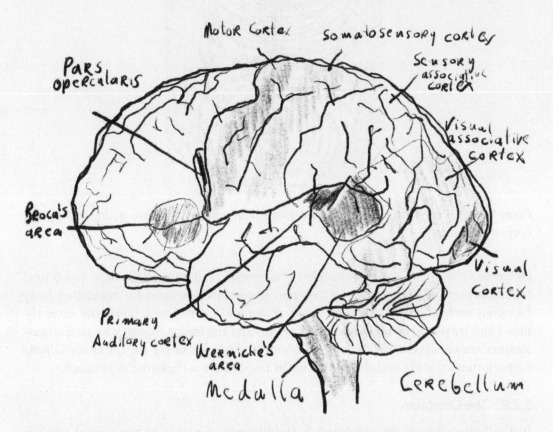

Figure 3.9 The motor cortex is the region of the cerebral cortex involved in the planning, control and execution of voluntary movements. Broca's area is a region in the frontal lobe of the dominant hemisphere, usually the left of the brain, with functions linked to speech production. Pars opercularis may be associated with recognizing a tone of voice in spoken native languages. Wernicke's area is one of the two parts of the cerebral cortex that are linked to speech, the other being Broca's area. It is involved in the comprehension of written and spoken language, in contrast to Broca's area, which is involved in the production of language. Broca's and Wernicke's areas are often found exclusively on the left hemisphere. Tactile representation is present at the somatosensory cortex. The association areas make decisions and send impulses through the motor areas. The visual cortex processes the visual information. The medulla contains the cardiac, respiratory, vomiting and vasomotor centers, and, therefore, deals with the autonomic functions of breathing, heart rate and blood pressure as well as the sleep wake cycle.

to skeletal muscles along nerve fibre passing down the brain stem and spinal cord. The human body is represented by an area of the motor cortex.

In anatomy, a lobe is a clear anatomical part. The four lobes of the brain were originally a purely anatomical classification, but have been shown also to be related to different brain functions [97, 184].

- The **frontal lobe** represented in the left part of the drawing (see Figure 3.9) includes the Broca's area and motor cortex. It is involved in strategic and effortful processing, goal driven behaviors. The motor cortex is the region of the cerebral cortex involved in the planning, control and execution of voluntary movements. Broca's area is a region in the frontal lobe of the dominant hemisphere, usually the left, of the brain with functions linked to speech production. Pars opercularis may be associated with recognizing a tone of voice in spoken native languages. Damage to the prefrontal cortex can result in issues with one's long term and short-term memories, as well as creates changes in people's behaviors and their abilities to plan and organize [110]. Phineas P. Gage (1823–1860) was an American railroad construction foreman who survival of an accident in which a large iron rod was driven completely through his head, destroying much of his left frontal lobe [179]. The injury had effect on his personality, such that his friends saw him not as the same person anymore.

- Wernicke's area is one of the two parts of the cerebral cortex that are linked to speech, the other being Broca's area. It is involved in the comprehension of written and spoken language. Broca's area is involved in the production of language. Broca's and Wernicke's areas are often found exclusively on the left hemisphere (laterization). The auditory cortex plays an important yet ambiguous role in hearing, neurons in the auditory cortex are organized according to the frequency of sound to which they respond best. They all form together the **temporal lob**.

- At the somatosensory cortex, tactile representation is orderly arranged from the toe to mouth. Each cerebral hemisphere of the primary somatosensory cortex only contains a tactile representation of the opposite side of the body. The density of cutaneous tactile receptors on a body part is generally indicative of the degree of sensitivity of tactile stimulation experienced. For this reason, the human lips and hands have a larger representation than other body parts. The association area make decisions and send impulses through the motor areas. They all form together the **parietal lobe**.

- The visual cortex processes the visual information and composes the **occipital lobe**.

- The **medulla** is a long stem-like structure, which makes up the lower part of the brainstem. The medulla contains the cardiac, respiratory, vomiting and vasomotor centers, and therefore deals with the autonomic functions of breathing, heart rate and blood pressure as well as the sleep wake cycle, see Figure 3.9.

3.3 BRAIN AS COMPUTER METAPHOR

3.3.1 Human Visual System

The human visual system is the most well understood system of the human brain [137, 139]. In biological vision the lens of each of the two eyes invert the image. The corresponding light representing it is captured in photoreceptors in the retina. Up is inverted down and right is represented on the left. This inverted image representation is learned to be corrected in the earliest childhood. The retina, which is a part of the brain, is covered with light-sensitive receptors the rods and the cons. Rods are primarily for night vision and to recognize movement. They are sensitive to broad spectrum of light, they sense intensity or shades of gray and can't discriminate between colors. Less common are cones, they are less sensitive and are used to sense color in day light vision. Center of retina has most of the cones (color) for high acuity of objects focused at center. The edge of retina is dominated by rods (shades of gray) that allows detecting motion of threats in the periphery. In the cones photopigments are used to sense color. There are three types: blue, green and red (really yellow) each sensitive to different band of light spectrum. Other colors are perceived by combining stimulation of the three colors. Photopigments are not evenly distributed, there are mainly reds (64%), very few blues (4%). The center of retina has no blue cones, that is why small blue objects disappear when fixated.

Perceptual information is processed in both hemispheres from the right retinas (left visual field) of both eyes and from the left retinas (right visual field) travel through the optic nerve, optic chiasm where the optic nerves cross to the optic tract. The optic tract carries retinal information relating to the whole visual field. The left optic tract corresponds to the right visual field, while the right optic tract corresponds to the left visual field [137].

The visual area of thalamus is the gateway through which visual information reaches the cerebral cortex. Visual processing includes spatial and temporal influences on visual signals that serve to adjust response gain, transform the temporal structure of retinal activity patterns and increase the signal-to-noise ratio of the retinal signal while preserving its basic content.

Sensory input originating from the left retinas (right visual field) travel to the left hemisphere of the primary visual cortex, also known as visual area 1 (V1), and from the right retina (left visual field) to the right hemisphere of the primary visual cortex, see Figure 3.10. The primary visual cortex (V1) is composed by neurons that receive connections from the retina and is divided into six functionally distinct layers. In the first layer, each neuron is connected to a subset of the retina corresponding to a local view of the input. The experimental work of Hubel and Wiesel [138], [139] that was performed first in the cat and later in the macaque monkey [139] indicated that the V1 is composed by two main types of cells [137], the simple cells and the complex cells. Simple cells react to stimuli with specific orientations and positions. Whereas complex cells react to stimuli with specific orientations but allow for positional shifts. In later stages, the complexity of the preferred stimuli of neurons increases [137] and at the end of the ventral stream, the cells are tuned to highly complex stimuli such as faces [116].

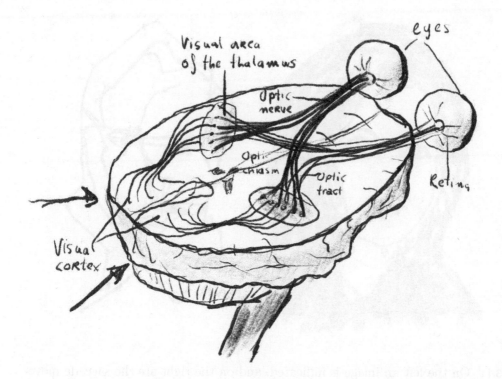

Figure 3.10 Perceptual information is processed in both hemispheres from the right retinas (left visual field) of both eyes and from the left retinas (right visual field) travel through the optic nerve, optic chiasm where the optic nerves cross to the optic tract. The optic tract carries retinal information relating to the whole visual field. The left optic tract corresponds to the right visual field, while the right optic tract corresponds to the left visual field. The visual area of thalamus is the gateway through which visual information reaches the cerebral cortex. Visual processing includes spatial and temporal influences on visual signals that serve to adjust response gain, transform the temporal structure of retinal activity patterns and increase the signal-to-noise ratio of the retinal signal while preserving its basic content. Sensory input originating from the left retinas (right visual field) travel to the left hemisphere of the primary visual cortex, also known as visual area 1 (V1), and from the right retina (left visual field) to the right hemisphere of the primary visual cortex.

It was suggested [116] that the brain includes two mechanisms for visual categorization [223]: one for the representation of the object and the other for the representation of the localization [159].

- The first mechanism is called the *what* pathway or ventral stream. It begins with V1, goes through visual area V2, to the temporal lobe, it is located at the lower part of the cortex. Visual area V2, or secondary visual cortex is the second major area in the visual cortex, and the first region within the visual association area.

Figure 3.11 On the left an image is indicated, and on the right are the saccade movements of the eyes. Object recognition results from fast eye movements, the saccadic movements in which points are fixated at a point of interest for 200–600 msec saccade (eye movement) of 20–100 msec to the next point of interest (points in the graph).

- The second mechanism is called the *where* pathway or the dorsal stream. It begins with V1, goes through visual area V2 to the parietal lobe, it is located at the upper part of the cortex. It is associated with motion, representation of object locations, and control of the eyes and arms, especially when visual information is used to guide saccades or reaching.

According to this division, the identity of a visual object can be coded apart from its location.

Object recognition results from fast eye movements, the saccadic movements in which points are fixated at a point of interest for 200–600 msec and a saccade (eye movement) of 20–100 msec to the next point of interest, see Figure 3.11. Point of interest correspond to changes in curvature of an object [226]. Attneave showed that a picture of a cat can be simplified by replacing all lines of low curvature with straight lines without adversely affecting the recognizability of the cat [226], see Figure 3.12. The lines of low curvature represent redundant information and can be later interpolated. Object recognition in the brain is based on temporal processing and not on a hierarchical template matching as used by convolutional neural networks.

The dorsal and ventral pathways from the retina to the higher levels of the visual cortex are separate at the lower levels of the visual system. At higher levels, they show increasing overlap. At lower level, we have an independent invariant processing

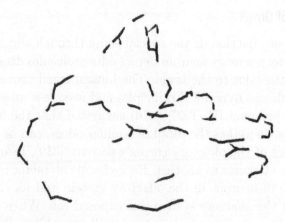

Figure 3.12 Attneave showed that a picture of a cat can be simplified by replacing all lines of low curvature with straight lines without adversely affecting the recognizability of the cat.

for colors, form, position motion and depth that are later integrated. There is still much controversy over the higher levels.

The visual representation is based on interpolation and the reconstruction of the reality. This becomes obvious during optical illusions or mental imaginary. Optical illusions are caused by the visual system and characterized by a visual percept that arguably appears to differ from reality.

Mental imagery is the mental invention or recreation of an experience that at least in some respects resembles the experience of actually perceiving an object in the absence of direct sensory stimulation [159]. The imagery is formed without perception through the construction of the represented object from memory.

The human visual system is algorithmic. There are already approaches in machine learning that go in this direction.

3.3.2 Human Acoustic Processing

The auditory pathway is well less understood but its working principles are roughly related to the principles of visual system. The main difference is that sound is purely mechanical and the auditory receptors are hair cells in the inner ear [167]. Acoustic energy, in the form of sound waves, is channeled into the ear canal by the outer ear. Sound waves strike the tympanic membrane, causing it to vibrate like a drum, and changing it into mechanical energy. The middle ear components mechanically amplifies sounds. The fluid movement within the cochlea in the inner ear causes membranes to shear against the hair cells. This creates an electrical signal, which is sent via the auditory nerve to the temporal lob, where sound is interpreted. The hair cells do not fire at the instant at which sounds begin or ends and an adaptation occurs to longer duration sounds.

3.3.3 The Sense of Smell

The sense of smell, or olfaction, is the special sense through which smells (or odors) are perceived. Olfactory sensory neurons detect odor molecules dissolved and transmit information about the odor to the brain. The human smell processing is an archaic system based on different dynamical principles and even less understood.

Walter Jackson Freeman III (1927–2016) suggested that the learning and recognition of novel odors, as well as the recall of familiar odors, can be explained through the chaotic dynamics of the olfactory cortex's activity [319]. Learned odors popped the system from one attractor to another. Each chaotic attractor represents the firing of particular groups of neurons in the olfactory system and its shape changed with each new scent that the olfactory system was exposed too. When a scent is detected and the olfactory system is alerted, these nerve cell assemblies allow instant recognition of familiar scents [257]. On smelling, a transition occurs from low level chaos to a trajectory, which in the case of a familiar odor will settle in one of several periodic orbits. In the case of a new odor, the existing periodic attractors move chaotically until a new periodic attractor is established. The way the brain uses chaos to ensure continual access to previously learned patterns is to develop these attractors for different learned inputs. The background chaotic activity enables the system to jump rapidly into one of these attractors when presented with the appropriate input.

3.3.4 Assembly Theory

The neural assembly theory was introduced by Donald Hebb [130]. He proposed a connection between the structures found in the nervous system and those involved in high-level cognition such as problem solving. An assembly of neurons acts as a closed system, and can therefore represent a complex object. Activation of some neurons of the assembly leads to the activation of the entire assembly, so that manipulation on the representation of a complex objects can be performed [131, 207]. The self-organized behavior means that the images we see in our minds are the collective activity of the mass action of thousands to millions of neurons spread over the brain, not the activity of single neurons [319].

The pump of thought model [40, 42, 206] is a theoretical assembly model that explains how thoughts represented by assemblies can be propagated and changed by the human brain. The process of human problem solving is described by this model as the transformation of thoughts through a sequence of assemblies [41, 206]. This description is related to the production system model.

3.4 NEUROMORPHIC COMPUTING

Neuromorphic computing explores emergent properties of large-scale neural architectures and benefits of using neuromorphic hardware to simulate biological neural networks [280].

The biological processes of neurons and their synapses are complex, and thus very difficult to simulate on current generation of digital computers. The characteristics of neurons and their coding can be abstracted into mathematical functions that closely

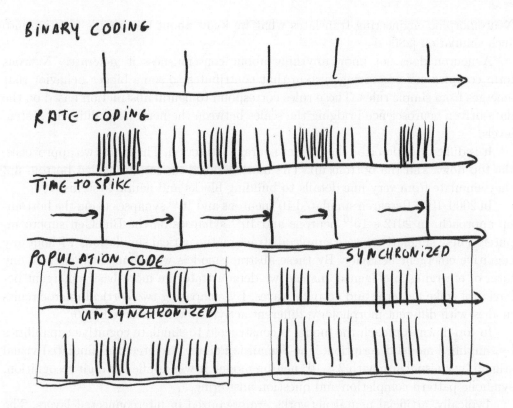

Figure 3.13 Encoding represented by spikes can be based on binary coding indicating when something happened, rate coding representing information as a function of spike rate, time to spike as a function of exact time at which neuron fired, and with population coding [280].

capture the essence of the operation. However, neural coding in cases with nonlinear feedback and non-deterministic generation of spikes is far from being understood [92].

Neural coding represented by spikes can be represented by rate coding, population coding or information as a dynamic pattern of firing, see Figure 3.13.

To simulate a spiking neuron, we must specify details like the action potential and the spike propagation. This is usually done by a model of neural dynamics like the Leaky Integrate-and-Fire model or the Izhikevich and Hodgkin-Huxley model [280]. Both models simulate the spikes behavior of the neurons. Neuromorphic computing does not perfectly mimic the brain and all its functions but simulates what is known of its structure and operations. Such simulation requires large computational power. However, there are some limits of integrated circuits, we cannot pack more transistors on a chip and cannot use higher clock rates. A solution is to change the computing paradigm leading to brain inspired neuromorphic hardware that goes around the memory CPU communication bottleneck. Such neuromorphic hardware supports the simulation of large networks. This can be done with neuromorphic device that uses physical artificial neurons (made from silicon) to do computations. To perform such simulations, we must understand the morphology (structure) of individual neurons, circuits, applications and overall architectures to creates desirable computations.

Neuromorphic engineering translates what we know about the brain's function into such simulation [280].

A neuron does not know anything about consciousness it generates. Neurons form complex self-organizing system that contributes to some bigger behavior that emerges from simple rules. These rules correspond to neural abstraction based on the data-driven neuroscience bridging the scales between the neurons at different system levels.

It is difficult to describe the correct neural abstraction. There are two approaches, the top down and the bottom up. The most detailed models are called bottom up, they simulate from very fine details to building blocks and neural systems [280].

In 2009, IBM Research simulated 10^9 neurons and 10^{13} synapses using the bottom-up approach. In 2012 s 10^{10} neurons and 10^{14} synapses on the BlueGen supercomputer. These simulations were simple abstract mathematical simulations without any cognitive emergent behavior. By these abstract models we cannot demonstrate any form of emerging intelligence, instead we demonstrate the more basic emergent behavior of spike patterns and activity waves. By comparing two cortical microcircuits models with different morphology, different activities were observed [280].

In top-down approach the model is constructed to simulate cognitive capabilities by simplified model description. The Semantic Pointer Architecture simulated visual stimuli by a visual arm with $2.5 \cdot 10^6$ neuron model. It was applied to digit recognition, symbolic pattern completion and question answering.

Typically, artificial neural networks are organized in interconnected layers. The brain's neural architecture is organized differently. There is no direct mapping between layers, there is no precise forward mapping even in the layered cortex structure. The architecture is based on cerebral columns that have neurons with corresponding characteristic that form functional structural entities. Columns are interconnected and create structural networks [280]. The neuromorphic computing explanation of mind is based on emergence. Emergence explains the human mind build on our ignorance of how local interaction produces emergent behavior. Another approach is based on living cells, the brain organoids. Brain organoids are self-organized stem cell of living human neurons. One tries to build brain models using organoids. However, they are hard to maintain, and it is difficult to study them at large scale [280].

3.5 NEUROIMAGING

To understand the brain, we need to understand functions of its building blocks. Neuroimaging techniques functional like magnetic resonance imaging (fMRI) and brain positron emission tomography (PET) have an enormous impact on the field of brain research. The neuroimaging methods PET and fMRI cannot access neuronal activity directly. They measure local properties of the cerebral blood flow [259].

Functional magnetic resonance imaging measures local properties of the cerebral blood flow and is based on blood oxygen level dependence. The brain activity is correlated with changes of activity associated with various stimulus conditions [54], see Figure 3.14.

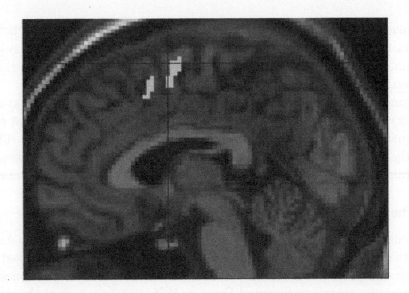

Figure 3.14 Changes of activity associated with various stimulus conditions are correlated with the brain activity. A cluster indicates a brain activity during an experiment (indicated by two gray blobs) [295].

PET measures emissions from radioactively chemicals that have been injected into the bloodstream [259].

Both techniques allow to describe the spatio-temporal neuronal activity processes in the working brain of humans with some constraints [259]. The spatial resolution of fMRI is the best achievable in whole brain imaging, however, it is still several orders of magnitude away from representing single neurons. The time resolution of fMRI is also constraint, it is from around 1 to 7 sec [295].

PET and fMRI are based on two hypothesizes, that the different brain regions are engaged in different brain functions and that the functions are formed by networks of interacting brain regions. Brain functions are formed by networks of interacting brain regions, different functions correspond to different functional networks. By measuring a covariance between different brain regions one can guess the function of such functional networks [295].

However after Kosslyn, "*It is unclear that we will come to a better understanding of mental processes simply by observing which neural loci are active while subjects perform a task*" [158]. According to him only experiments on animals (see Figure 3.1), known effects of brain damage on human behavior and cognition, or facts from neuroanatomy, like connections between areas can lead to understand the working principles of the human brain.

3.6 CONSCIOUSNESS AND THE BRAIN

What is the difference between one simulation of the brain on a running on a computer, or two such simulations running in parallel on the same computer? How many minds can a human have?

It makes sense to have both automated behavioral programs that can be executed rapidly in a stereotyped and automated manner, and a slightly slower system that allows time for thinking and planning of more complex behavior. Does the slightly slower system corresponds to our consciousness?

It is unclear what adaptive **advantage** consciousness could provide at all. Some researchers assume that consciousness did not evolve as an adaptation but arrived as a consequence of other developments such as increases in the brain size [154]. They assume that consciousness is a computational system that is realized by some regions of the brain of increased size, an emergent property of these parts of the brain.

3.6.1 The Unconscious

The unconscious consists of the processes in the mind, which occur automatically and of each the mind is not aware. Unconscious phenomena include repressed feelings, automatic skills, subliminal perceptions and automatic reactions. The concept of the unconscious was central to Sigmund Freud's (1856–1939) account of the mind [101]. Unconscious can be seen as automatic thoughts and the locus of implicit knowledge, things that we have learned so well that we do them without thinking.

French polymath and pioneering mathematician Henri Poincaré (1854–1912) described the unconscious process of invention, he observed a process profoundly applicable not only to mathematics, but to just about any creative discipline [218].

Poincaré wanted to represent some functions by the quotient of two series in analogy with elliptic functions. He asked himself what properties these series must have if they existed. He succeeded without difficulty in forming the series by some functions. Then he went to a geologic excursion and forgot his mathematical problem, later when he entered an omnibus the idea came back to him with the solution. It became clear to him that the transformations he had used to define the functions were identical with those of non Euclidian geometry.

Poincare ensures this is a pattern rather than mere one-off coincidence by citing another example. He turned his attention to the study of some arithmetical questions without much success and without a suspicion of any connection with his preceding researches. Disgusted with his failure, he spend few days at the seaside and thought of something else. One morning, walking on the bluff, the solution came to him with and immediate certainty.

The sudden illumination appears after long unconscious prior work but only if one hand preceded by a period of conscious work. It only happens after some days of voluntary effort without any results.

Either the unconscious is searching computational for a solution in the background without the knowledge of the mind, or it is capable of contact with this eternal world of ideas as suggested by Roger Penrose [212].

3.6.2 Global Workspace Theory

Global workspace theory (GWT) is a simple cognitive architecture that can be explained in terms of a "theater metaphor." In the "theater of consciousness" the stage corresponds to the contents of consciousness, like some actors on the stage. The audience is watching the play. Behind the scenes there are also some other workers that contribute to the play, like script writers, scene designers that correspond to unconsciousness process [17, 18].

The global workspace is closely related to conscious experience, though not identical to it, it is distinct from the concept of the Cartesian theater, since it is not based on the implicit dualistic assumption of "someone" viewing the theater.

Consciousness arises when information that is processed in unconscious modules is broadcast to a central working memory. The working memory (or short term memory) involves a fleeting memory with a duration of a few seconds. Neural activity probably via global feedback from frontal regions of neocortex back to sensory cortical areas and related structures supports the working memory (short-term memory). Competition prevents more than one or a very small number of percepts to be simultaneously and actively represented. The contents correspond to what we are conscious of, in contrast to a multitude of unconscious cognitive brain processes.

This approach is highly related to the production system model, like the SOAR.

Susan Blackmore challenged the concept of stream of consciousness [35], by stating that "*When I say that consciousness is an illusion I do not mean that consciousness does not exist. I mean that consciousness is not what it appears to be. If it seems to be a continuous stream of rich and detailed experiences, happening one after the other to a conscious person, this is the illusion.*"

The continuity of the "stream of consciousness" may in fact be illusory, just as the continuity of a movie is illusory.

3.6.3 Searchlight Theory

It is supposed that the brain has an internal attentional search light that corresponds to the visual contents of consciousness [88]. Attention is linked to a search light that is focused on the cued location and shifted as necessary [159, 223]. The search light moves around from one visual object to the next with steps as big as 70 ms and fulfills following criteria. It has been able to sample the activity in the visual buffer (working memory) and decide, which object is important. After this, it moves to the next place demanding attention and repeat the process.

Crick proposed that the search light is controlled by the thalamus [69]. In his view, the expression of the search light is the production of rapid firing in a subset of active thalamus neurons and the corresponding fields in the visual cortex.

All visual input to the cortex passes through the thalamus, and there is a rapid movement of the searchlight from place to place. The brain must know what it is searching for. That aspect involves the parietal and the temporal lobes. The visual buffer (working memory) consists of the object and coordinate spatial encoding in parietal and temporal lobe and of an image representation in the occipital lobe.

3.6.4 Correlation and Consciousness

One can avoid philosophical debates that are associated with the study of consciousness, by emphasizing the search for "correlation" and not "causation" [154].

The Neuronal Correlates of Consciousness (NCC) is represented by the smallest set of neural events and structures sufficient for a given conscious experience. The set should be minimal to be able to understand, which parts of the brain are necessary to produce it.

To be conscious of anything the brain must be in a relatively high state of arousal. Different levels or states of consciousness correspond to different kinds of conscious experiences. The wakefulness state is quite different from the REM sleep, dreams that are usually not remembered. Altered states of consciousness are present when perception and insight may be enhanced compared to the normal waking state.

Neuroscientists use empirical approaches to discover neural correlates of subjective phenomena. Cortical areas are analyzed separately. Their contribution to consciousness is explained by synchronized action potentials in neocortical pyramidal neurons. Synchronized action potentials means that the frequency of the firing neurons is synchronized, the same frequency with the same phase.

A certain neural activity represents qualia, that according to Christof Koch are mental shorthand for the mental states [154, 155].

3.6.4.1 Reductionism and Qualia

Eliminative reductionism, at its core, subsumes that all mental phenomena can be explained by the functioning of their neurological correlates. Colors, sounds and smells do not exist in the outside world. They are the creations of our brain in response to light waves, rhythmic variations of air pressure. Qualia (singular quale) are defined as individual instances of colors, sounds and smell. Qualia an emerging neural activity of brain activity triggered by external and internal processes. For example, Daniel Dennett argue that qualia do not exist and are incompatible with neuroscience.

On the other hand, António Damásio argues that qualia exist and have causal efficacy [75] and Penrose [126, 214] assumes that the corresponding physical laws describing qualia have yet to be discovered.

3.6.5 Split-Brain

The two hemispheres of the cerebral cortex are linked by the corpus callosum, through which they communicate.

After the right and left brain are separated, each hemisphere will have its own separate perception, concepts and impulses to act.

When split-brain patients are shown an image only in the left half of each eye's visual field, they cannot vocally name what they have seen. This is because the image seen in the left visual field is sent only to the right side of the brain, and most people's speech-control center is on the left side of the brain. The patient cannot say out loud the name of what the right side of the brain is seeing since the communication between the visual processing and speech-control is not possible [64].

If a split-brain patient is shown a picture of a chicken foot and a snowy field in separate visual fields, both pictures a processed individually. If the patient is asked to relate words from a list that best relate with the pictures, she would choose a chicken to associate with the chicken foot and a shovel to associate with the snow. If asked why she has chosen the shovel, she would replay to relate to the chicken. It seems that the choices were made unaware of each other, and the explanation is performed by one hemisphere. A split-brain subject is composed of two conscious psychological beings that fail to recognize each other's existence and indeed cannot distinguish themselves from each other. One could assume that a split-brain subject has two minds. However, it seems that split-brain subject is one in the psychological sense, she identifies herself as whole, as one person [242].

3.6.6 Libet's Experiment

In the 1980s, Benjamin Libet asked each subject to choose a random moment to flick their wrist while he measured the associated activity in their brain, the activity in the motor cortex and supplementary motor area of the brain leading up to voluntary muscle movement. He found that the activity started about 0.35 sec earlier than the subject's reported conscious awareness "now" he or she feels the desire to make a movement. Libet concludes that we have no free will in the initiation of our movements [173].

The results suggest that decisions made by a subject are first being made on a subconscious level before one is consciously aware of having made it. The person is only under an illusion of having made the decision since the consciousness could not initiate the behavior. The causal role of consciousness seems not to exist.

In another Libet's experiment the patients' somatosensory cortices were electrically stimulated during surgery, 0.5 sec of continuous stimulation before the patient had any perception. Our body responds before we are conscious of why it is responding, after neuronal adequacy is reached, the event is referred back to the point at which it occurred. Libet's conclusions was that backward referral challenges materialism and supports dualism, since consciousness would not equate certain brain activity.

Susan Blackmore [35] states that *"conscious experience takes some time to build up and is much too slow to be responsible for making things happen."*

3.6.7 Temporal Window

One of the cognitive brain functions is to provide a causally consistent explanation of events to maintain self-identity over time, leading to the psychological concept of "now." Identity is a concept that defines the properties of a rational person over time. It is a unifying concept based on the biological principles of homeostasis [32], [115]. Organisms must maintain stability, e.g., the regulation of body temperature, to guarantee the maintenance of life. This principle is extended by allostasis [262] for the regulation of bodily functions over time. To perform this task, efficient mechanisms for the prediction of future states are necessary to anticipate future environmental constellations [21, 285].

This is done, because the homeostatic state may be violated by unexpected changes in the future. It means as well that every organism implies a kind of self

Figure 3.15 Ernst Pöppel, a German psychologist and neuroscientist, suggested that in humans a temporal window with the duration of 3 sec is created. This window represents the psychological concept of "now."

identity over time [322]. This identity requires a time interval of finite duration within which sensory information is integrated. Different sensor information arrives at different time stamps. The fusion process has to be done over some time window. Similar problems are present during a sensor fusion task in a mobile robot. For example, in visual and auditory perception in humans the transduction of the acoustic information is much shorter than the visual [219]. Ernst Pöppel suggested that in humans a temporal window with the duration of 3 sec is created [220], see Figure 3.15. This window represents the psychological concept of "now" [322]. The consciousness concept of "now" represented by the temporal window is shifted backward in time of the consciousness itself, since a subconsciousness mechanism is required to perform the integration task. The consciousness is present in the "now," it recovers the information what was before with the aid of information that is stored in the memory and tries to predict what will happen later.

3.6.7.1 Cartesian Doubt

Rene Descartes supposed that there existed nothing [80], However, he finds that it is impossible to doubt that he himself exists. The act of doubting one's own existence served as proof of the reality of one's own mind; thinking is occurring in a moment, I do not know what happened before or what will happen after.

3.7 DREAMS AND MIND

3.7.1 Conscious Experiences

A dream is composed of images, thoughts, sensations and emotions that are experienced by mind during certain stages of sleep. The early writings from writings from early Mediterranean civilizations indicate the nature of dreaming changed between Bronze Age antiquity and the beginnings of the modern times. In the antiquity dreams were composed of passive hearing and visitation. In modern times, the dreamer become a character who actively participates in the dreams. Aristotle [12] already noted that one can sometimes be aware while dreaming that one is dreaming. If we assume that dreams are experiences, we do not know how exactly to characterize dreaming relative to wake-state psychological terms. A reason to believe that dreams are experiences during sleep is the relationship between dreaming and rapid eye movement (REM) sleep. Sleep is not a uniform state of rest and passivity, but there are different stages of sleep [14, 15, 78]. Following sleep onset, periods of non REM are followed by periods of REM sleep. Dreams have the same character as waking perception, they put us in contact with mind-independent objects. The actual conscious experiencing of dreams seems to be similar to that proposed for waking consciousness.

If we assume that dreams are conscious experiences, we can become sceptic about sensory-based beliefs about the external world and his own bodily existence. We could assume that we have a lifelong dream and none of our experiences have ever been caused by external objects. We could also assume that we sometimes dream but cannot rule out at any given moment that we dream right now.

3.7.2 Cognition

Dreams are a global state of consciousness in which experience arises under altered neurophysiological conditions as compared to standard wakefulness. Dreams occur spontaneously and regularly. We could regard dreams as a test case for theories of consciousness [61,62]. Dreams have also been suggested as a test case for whether consciousness is independent from cognitive access [63]. Dreams could provide empirical evidence that conscious experience can occur independently of cognitive access [246]. This is because during REM-sleep dreams, the dorsolateral prefrontal cortex as the most plausible mechanism underlying cognitive access is selectively deactivated.

3.7.3 Dreams and Identity

We have a self in dreams, this self can sometimes be a slightly different (e.g. older or younger) version of our waking self or even a different person. Dreams therefore raise interesting questions about the identity between the dream and waking self. The personal identity depends on psychological continuity, including memory recall. In the absence of memory recall, the identity cannot be maintained [152].

Dreams in which the protagonist of the dream seems to be a different person from the dreamer, are particularly difficult to understand with respect to identity. It seems that the identity is autobiographically defined over a time window and that this identity may be replaced. The dream self is at the center of the dreamed events, which define d the experience of a self in a world. This leads to further questions about the self-experience in dreams and how it is different from waking self-experience. The consciousness experience in a dream is as reading to us when we read a story or watch a film.

3.7.4 The Meaning

We still do understand the purpose of sleep and dreams, we also do not know the functions of REM (rapid eye movement) sleep. Theories on the function of REM sleep and dreaming contend that their function is probably a combination of following points:

- **Consolidate memories:** According to the information-processing theory, sleep allows us to consolidate and process all of the information and memories that we have collected during the previous day. It is even suggested that dreaming is an active part of learning, mathematically formulated by the Helmholtz machine [76]. On the other hand, the reverse-learning theory suggests that we dream to forget. Important memories are stored and unimportant memories are erased. Most dreams are forgotten, it was proposed [70] that REM sleep "erases" or deletes unnecessary memories. A computer metaphor suggest that dreams serve to "clean up" clutter from the mind, refreshing the brain for the next day. These assumptions are compatible with the computationalism that states that the human mind is an information processing system, and that cognition and consciousness together are a form of computation. In this sense, androids do dream of electric sheep.

- **Process emotions:** Under the continuity hypothesis, dreams function as a reflection of a person's real life, incorporating conscious experiences into their dreams. Rather than a straightforward replay of waking life, dreams show up as a patchwork of memory fragments. Dreams express our deepest desires. Sigmund Freud called dream interpretation the "royal road" to the unconscious [102]. In Freud's view, dreams are formed as the result of two mental processes. The first process involves unconscious forces that construct a wish that is expressed by the dream, and the second is the process of censorship that forcibly distorts the expression of the wish. Every dream has a connection point with an experience of the previous day. Though, the connection may be minor, as the dream content can be selected from any part of the previous time.

- **Gain practice confronting potential dangers:** During dreaming amygdala is one of the most active areas of the brain, amygdala is the part of the brain associated with the survival instinct and the fight-or-flight response. Because the amygdala is more active during sleep than in your waking life, it may be the brain's way to prepare us for life-threatening and emotionally intense scenarios including training our fight-or-flight instincts.

- **Gain practice in constructing a narrative:** As described by "When Brains Dream" [320] dreaming serves as the ability to imagine possibilities, evaluate

Figure 3.16 António Rosa Damásio, a Portuguese-American neuroscientist, suggested that the self is the key to conscious minds.

them, and thereby plan future actions. This was as well proposed by António Damásio, in his 2000 book "The Feeling of What Happens" [74]. In an altered state of consciousness during dreaming imagined narratives are constructed resulting in emotions. Associations between recently formed memories usually from the preceding day are related to older memories, and a narrative it constructs that induces an emotional response in the brain. The narrative construction during dreaming is based on associations of recent events. Dreams are disconnected narratives that are not integrated in our autobiographic narrative, they are forgotten and disappear. Dream allow us to explore different new possibilities, they show us what has been and what might be. They serve as a preparation for future storytelling.

3.8 DAMÁSIO'S THEORY OF CONSCIOUSNESS

António Damásio, a Portuguese-American neuroscientist, argues in his 2000 book "The Feeling of What Happens" [74] that consciousness provides two functions: To construct narratives and to feel one's emotional response to them. They give humans and other conscious animals the ability to imagine possibilities, evaluate them and to plan future actions.

The consciousness is based on a hierarchy of three stages with each stage building upon the last. The most basic representation is the Protoself, next is Core Consciousness, and finally, Extended Consciousness.

The Protoself is a non conscious state, it is the most basic level of awareness signified by a collection of neural patterns that are representative of the body's internal state. The function of Protoself is to constantly detect and record, moment by moment, the internal physical changes that affect the homeostasis of the organism.

Brain areas like the hypothalamus, which controls the general homeostasis of the organism, the brain stem, whose nuclei map body signals, and the insular cortex whose function is linked to emotion work together to keep up with the constant process of collecting neural patterns to map the current status of the body's responses to environmental changes.

The Core Consciousness occurs when an organism becomes consciously aware of feelings associated with changes occurring too its internal bodily state. The Core Consciousness can recognize that its thoughts are its own. It builds a momentary sense of self, based on communications received from the Protoself. A relationship is established, between the organism and the object it is observing as the brain continuously creates images to represent the organism's experience of qualia. Feeling' emerges as a still unconscious state which simply senses the changes affecting the Protoself due to the emotional state, it is the feeling of knowing a feeling. Core consciousness is concerned only with the present moment, here and now. It does not require language or memory, nor can it reflect on past experiences or project itself into the future.

When consciousness moves beyond the here and now, Damasio's third and final layer emerges as Extended Consciousness. The Extended Consciousness requires memory to go backward in time. The autobiographical self draws on memory of past experiences and involves use of cognitive mechanism that project itself into the future. Over time the autobiographical self develops, the working memory is necessary for an extensive display of items by mental images to be recalled from the long-term memory.

3.9 THE EXPLANATION HYPOTHESIS

The explanation hypothesis is related to Damásio's theory of consciousness; however, it relies on a temporal window. Split brain research and the stimulation of brain regions during awake operations suggest that the brain generates an explanation of effects that are not initiated by consciousness [173], [64]. Before an event occurs, an explanation must be incited by the subconscious parts of the brain, so the explanation can be integrated into the temporal window of the self when the event occurs [304, 306]. Additionally, other organic functions must be notified due to some predicted possible events. If an explanation is not possible due to a lack of causality, the identity of the self may fracture. The implication is that personal identities can only exist in meaningful causal worlds. This concept is related to general constructor theory, posited by David Deutsch [86]. In this context, the mind defines the psychological concept of "now." The identity is more related to the human senses than it is to an algorithmic device. From this perspective, artificial intelligence models human cognition but not the mind or consciousness. The main task of the brain is to provide a causally consistent explanation of events and to decide which actions will maintain the homeostatic state. The mind facilitates the maintenance of self-identity over time that leads to an evolutionary advantage. The mind is needed to help to make adaptive predictions in an unexpected, changing future that cannot be described by rules or heuristics. Self-identity and self-consciousness correspond to a combination of qualia that indicate the quality of a causal story. Through the reward system, we search for a combination that induces our emotional system. We feel the story, we sense the story through the consciousness, which acts as a kind of a perceptive organ. Like the conscious experience of the taste of wine or music, we experience the beauty of a story. For this reason, we like to read books or see films: They act as a kind of drug that triggers our reward system in the absence of our own story. In this highly restricted view, qualia are nonphysical and are perceived by our minds and incorporated into the experience of self-consciousness.

3.10 OPEN QUESTIONS

Many questions are still not fully understood, why does the illusion of motion result when a series of film images are displayed in quick succession? What are the physical laws that describe qualia? We could answer this question with the argument that qualia do not exist, they are an illusion, a distortion. But this argument is like to tell to someone who has teeth pain, look it is just an illusion, ignore the pain! If we give the person some drugs, the pain disappears. After all it was not an illusion.

One of important questions in our life is if the free will is an illusion? The deterministic universe is logically incompatible with the notion that people have free will, but then again, it is supposed that nondeterministic quantum mechanics plays an essential role in the understanding of the human mind and free will.

CHAPTER 4

Quantum Reality

THE Second law of thermodynamic states that isolated systems tend toward an equilibrium macrostate with as large total entropy as possible corresponding to the largest number of microstates.

4.1 ENTROPY AND INFORMATION

A thermodynamic description is not a description of a physical property and relationships, but rather a fundamental property. Thermodynamic relations might remain valid for different physical theories. They link physical entities to their organization and to their informational structure. Thermodynamics developed out of a desire to increase the efficiency of early steam engines, see Figure 4.1.

Thermodynamics is a branch of physics that deals with heat, work and temperature, and their relation to energy, radiation, and physical properties of matter. It was James Clerk Maxwell, (1831–1879), a Scottish mathematician, who was one of the first to recognize the connection between thermodynamic quantities that are associated with gas, such as temperature, and the statistical descriptions of the molecules [185]. The behavior of gas on a macroscopic (big) scale like temperature, pressure and volume uses statistics to describe the microscopic (small) states of the molecules. We cannot specify the exact speed or position of each molecule, instead we describe its statistical behavior.

Macroscopic (large scale) properties such as density or pressure are the result of microscopic (small scale) properties. A macroscopic property can be realized by different microscopic states. Microscopic states are not static, but they continuously change corresponding to the motion of atoms or molecules. Statistical mechanics describes this motion by some random parameters. From the macroscopic viewpoint, each molecule is moving randomly. A set of molecules is described by a distribution of random variables that express these random movements. *"A physical system that is made up of many, many tiny parts will have microscopic details to its physical behavior that are not easy to observe. There are various microscopic states the system can have, each of which is defined by the state of motion of every one of its atoms, for instance."* a quote from Matt McIrvin.[1]

---

[1] Matt McIrvin physics- group posting.

DOI: 10.1201/9781003244547-4

Figure 4.1 Thermodynamics developed out of a desire to increase the efficiency of early steam engines. An early English steam locomotive (1845) hauling a tender and three carriages adapted from horse-drawn coaches is shown.

4.1.1 Second Law of Thermodynamics

Suppose that we have two dice; the macrostate is the total of the two dice, and the microstate corresponds to the number on each of the die. There are six ways to get a total of 7 from the microstates of the two dice but only one way to get a total of 2 or 12. For this reason, throwing two dice to sum a total of seven is more likely to occur; a sum of seven is six times more likely to occur than two or twelve. Each die can have one out of six microstates. The number of microstates of the whole system of two dice is $6 \times 6 = 36$,

For n fair dice, the total number of possible microstates is at a maximum at a macrostate value of $3.5 \times n$ ($3.5 \times 2 = 7$). If we shake n dice in a bag and measure the uppermost faces, and we repeat the experiment, the dice will converge rapidly on the value (the "macrostate") for which the number of ways to make the value from individual dice ("microstates") is at a maximum.

By performing this action, we model the thermal fluctuations that describe the second law of thermodynamics. Thermal fluctuation is described by randomly uncorrelated movements of molecules. Randomness in this context means that there is no pattern or predictability in events.

The macroscopically measurable quantities converge to the values that have the largest number of microstates. The largest number of microstates models the entropy. Entropy is a measure of disorder of the configuration of the states of molecules or other particles, which make up the system, like a gas in a closed system. The second law states that the total entropy of any isolated thermodynamic system tends to increase over time, approaching a maximum value. It appears that the basis of the time direction is related to the second law of thermodynamics. The growth of entropy corresponds to the arrow of time.

The other two laws are the following, the first law states that the change in the internal energy of a closed thermodynamic system is equal to the sum of the amount of heat energy supplied to the system and the work performed on the system [305]. The third law states that if a system asymptotically approaches the absolute zero of the temperature, the entropy of the system asymptotically approaches a minimum value. All processes virtually cease, there are no thermal fluctuations.

The second law states that the total entropy of any isolated thermodynamic system tends to increase over time, approaching a maximum value, this is due to the thermal fluctuations that are modeled by random movements of the molecules.

4.2 INFORMATION

Suppose that we have two chambers that are separated by a common partition, which could be removed to permit the objects in one to move freely to the other. One chamber contains gas and the other chamber contains nothing; on the removal of the partition, the gas will rapidly diffuse and fill the empty chamber. Reverse evidence of this sequence mostly does not occur. This is because an isolated systems tends toward an equilibrium macrostate that corresponds to the largest number of microstates that are present when the gas is diffused over the two chambers. Thinking in terms of the dice model, the probability of reverse evidence approaches zero but is never equal to zero. If we put an ice cube in a lemonade, it will dissolve. After it dissolved it should not appear again, however the probability is not zero.

In the Maxwell paradox, there is a demon that observes the gas molecules. Between the chambers, there is a small door. The demon can open and close the small door, passing only one molecule into a chamber. When a molecule in its random motion heads toward the door of the chamber, it opens the door, briefly permitting the molecule to pass into the other chamber. Soon there will be more molecules in one chamber than the other. Because the demon requires almost no energy to operate the door, the process decreases the entropy, which is a contradiction to the second law of thermodynamics. In 1929, Leó Szilárd explained the thermodynamic Maxwell paradox by identifying information with the negative measure of entropy [266].

To perform this task, the demon must be very well informed about the position and velocity of the molecules that approached the door. Only with this information he can judge when and for how long the door should be opened to pass a molecule through the door into the chamber. He should not allow any molecules to pass in the opposite direction. As the demon's information about the distribution of the gas increases, the entropy of the gas decreases.

The negative of the demon's entropy increments as the measure of the quantity of the information that it has used. Information increments when entropy decrements. The information must be stored in a demon's memory. Given the fact that the memory is finite, the demon must erase some information. The erasing of information increases the entropy represented by energy. Bennett (1973) showed [26] that overwriting of one bit causes at least $k \cdot T \cdot log2$ joules of energy dissipation, where k is Boltzmann's constant and T is the absolute temperature. For example, the AND and OR gates erases one bit and generates energy dissipation (heat) during computation [166], [165]. Since $(1\ AND\ 0) = (0\ AND\ 1) = (0\ AND\ 0) = 0$, $(1\ OR\ 0) = (0\ AOR\ 1) = (1\ OR\ 1) = 1$ [27], having the result we do not know what the input was. On the other hand, a NOT gate is reversible in the sense that one can infer the output from the input. During the computation, information is lost and at the same time waste heat is generated [28,165]. For this reason, computer processors must be cooled to keep them within permissible operating temperature limits. We can convert information in energy, and energy in information. In this sense information is as well a part of the physical universe as matter and energy [263]. Information does not concern the substance and the forces of the physical world [263]. It is what remains after one abstract from the material aspects of physical reality.

4.2.1 Shannon's Entropy

Instead of a demon that operates a door between two chambers, let us imagine a simple experiment, for example, throwing a fair coin [277]. Before we perform the experiment, we do not know what the result will be; we are uncertain about the outcome. We measure the uncertainty by the entropy of the experiment and describe the experiment itself by probabilities. For example, for the outcome of the flip of an honest coin, the probability for a head or tail is 0.5. The probability is numerical degree of belief between 0 and 1 that provides a way of summarizing the uncertainty.

The experiment starts at t_0 and ends at t_1. At t_0, we have no information about the results of the experiment, and at t_1, we have all of the information, so that the entropy of the experiment is 0. This kind of information is called the Shannon's entropy. Manuela knows the outcome, but person Pedro does not. Shannon's entropy is defined by the number of most basic questions with yes or no answer Pedro could ask Manuela about the outcome of the experiment [250, 277]. A most basic question corresponds to the smallest information unit that could correspond to a yes or no answer. The smallest information unit is called a binary digit, or bit. The number of questions converge to a value that assumes that Manuela repeats the experiment an infinite number of times.

4.2.2 Information and Surprise

We say that events that seldom happen have a higher surprise than quite common event. For example, the event that a dog bites man is quite common. If we hear about it, we are not surprised. However, it is very seldom that a man bites a dog, if we hear it, we are truly surprised. We state as well that information is inversely related to probability [277].

4.3 RANDOMNESS

Where do the probability values come from and what is their relation to randomness? We can determine subjectively the numerical degree of belief. The true values can be determined from the true nature of the universe, for example, for a fair coin, the probability of heads is 0.5. Finally, for a finite sample we can estimate the true fraction by empirical experiments. We count the frequency of an event in a sample. This approach is called frequentist.

The toss of a fair coin with the probability of heads is 0.5 represented by 1 and tail 0 is indicated by a sequence

$$0, 1, 0, 1, 0, 1, 0, 1, 0, 1, 0, 1, 0, 1, 0, 1, 0, 1.$$

There is correlation between 0 and 1 represented by the repeating pattern $0, 1$, that is why the sequence is not random. In a sequence of random numbers there should be no correlation among the numbers in the sequence. A random number alone does not exist, randomness can be only defined by numbers in a sequence [313].

Classical physics is deterministic. Due of that, no true randomness exists in its context. Some facts could appear to be random. However, this arrangement is only the case because some essential information is missing.

A sequence could look random even though it is generated by a nonlinear deterministic equation like the pseudo random number generator.

Can people generate true random sequences? Hagelbarger asked subjects to create a binary random sequence, and the sequence was analyzed for correlations by a computer program. It turned out that humans could not generate true binary random sequences in the experiment [249].

In a deterministic universe no true randomness exist and in such a universe people have no true free will.

4.3.1 Laplace's Demon and the Cat

Let as assume our universe is deterministic. A Geiger counter measures the decay of a radioactive substance. There is a fifty percent chance that, in a given time frame, decay is measured. The Geiger counter is connected to a device that kills the cat, if decay is measured. Because the cat and the Geiger counter are in a closed room, we do not know whether the cat is dead or alive. Each of these possibilities is associated with a specific fifty percent probability. The cat is either dead or alive, but not at the same time. We do not know it since we have no access to the room. If someone (Laplace's demon) knows the precise location and momentum of every atom in this deterministic universe, their past and future values for any given time are entailed; he can calculate it by the laws of classical mechanics and determine if the cat is alive or dead, see Figure 4.2. The Laplace's demon will always remain infinitely removed from our human mind. The human mind has some knowledge about a system, but this knowledge does not describe the system. A map is a symbolic depiction emphasizing relationships between elements of some space, such as objects, regions or themes drawn to a scale expressed as a ratio of the real distance. A map can never represent

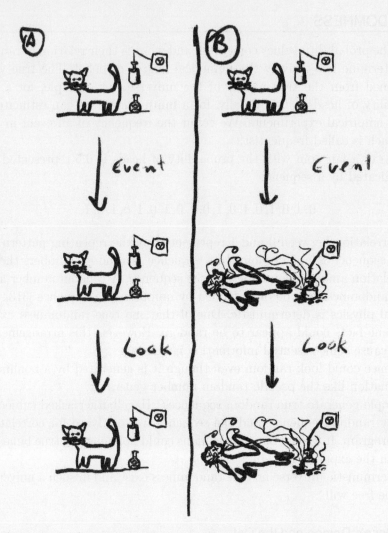

Figure 4.2 A Geiger counter measures the decay of a radioactive substance. There is a fifty percent chance that in a given time frame decay is measured. The Geiger counter is connected to a device that kills the cat if decay is measured. At some point the decay is measured or not; this corresponds to an event that happens or not. Because the cat and the Geiger counter are in a closed room, we do not know whether the cat is dead or alive. Each of these possibilities is associated with a specific fifty percent probability. The cat is either dead (A) or alive (B), but not at the same time. We do not know since we have no access to the room. If we look into the room the cat is either alive or dead. If someone (Laplace's demon) knows the precise location and momentum of every atom in this deterministic universe, their past and future values for any given time are entailed; he can calculate it by the laws of classical mechanics and determine if the cat is alive or dead.

the space correctly, to do this we would require a map of the same scale as the reality itself.

4.3.2 Broken Causal Chain

Is our universe deterministic? Only in a nondeterministic universe free will and true randomness can exist. A true random event cannot be an effect of a causal chain. A causal chain is an ordered sequence of events in which one event in the chain causes the next. True randomness can only originate outside the system itself, our universe. The same is true for a mind capable of free will. This view leads to the dualistic problem of how the mind outside the universe interacts with the deterministic universe itself. Yet we know through quantum physics that our universe is not deterministic.

4.4 QUANTUM EVOLUTION

Statistical laws govern the totality of observations in physics. An object can be described in classical mechanics by a vector which describes the position and it's momentum. Classical mechanics is usually valid at the macro scale. The changes in the position and the momentum of the object over time are described by the Hamiltonian equation of motion. The state itself of the object is described by the Hamiltonian function.

At micro scale the observations are described by quantum physics. A photon can be described by a wave function if it is isolated from its environment. An event is unobserved if it is isolated from us, we are unaware of the event and unable to get any information about it.

Light appears only in chunks that can be quantized. An individual chunk is called quantum, a light chunk is called a photon [83]. Quantum theory gets its name from this property, which it attributes to all measurable physical quantities. There are no measurable continuous in quantum physics. The notion of continuous range of possible values is just an idealization.

The wave function in quantum mechanics, if unobserved, evolves in a smooth and continuous way according to the Schrödinger equation, which is related to the Hamiltonian equation of motion. This equation describes a linear superposition of different states at time t. A general solution of the Schroedinger equation represents the unitary (linear) evolution that is deterministic and reversible. Reversible means that no information is lost. According to Susskind, this is the "Zero law" of physics [265].

The vector $\mathbf{x}(t)$ (for simplicity we will call it as well a wave) describes the probability of the presence of certain states at time t. A dimension represents each state, and the value of the vector is related to the probability of the state being present. This evolution is done in parallel over all states of the vector $\mathbf{x}(t)$.

During the observation (measurement by the observer, by us), the wave collapses into one definite state with a certain probability. This state corresponds to one dimension of the vector $\mathbf{x}(t)$. The measurements always find the physical system to be in a definite state. It does something to the wave function represented by the vector $\mathbf{x}(t)$. This something is not explained by quantum theory.

A quantum system that is perfectly isolated maintains its wave representation, its coherence. If it is not perfectly isolated, during interaction with the environment the wave representation of an event is lost, this is called the quantum decoherence. Quantum decoherence happens during the measurement.

Decoherence is usually viewed as the loss of information from a system into the environment. However, the quantum decoherence provides us only with a framework for apparent wave-function collapse, it does not explain the collapse itself. Since quantum systems are not isolated in a large scale (macro scale), the effects are mostly present at the scale of atoms, and subatomic particles (micro scale).

For instance, let us imagine a gun that fires electrons and a screen with two narrow slits x and y and a photographic plate. An emitted electron can pass through slit x or slit y. The electron detectors show from which slit the electron went through (measurement), and we find that the probability of the electron hitting the photographic plate is

$$p(x) + p(y).$$

This probability means that, when measured, the electron behaved as a particle.

On the other hand, if we remove the detectors, the electron is unobserved, not knowing through which slit it went through. Now, the electron is represented as a wave with the amplitudes, two waves pass at the same time through both slits. At different positions at the photographic plate, an interference pattern emerges due to the different phase that change with time. The probability of the electron hitting the photographic plate is

$$p(x) + p(y) + Interference$$

with an interference that can be positive, not present, or negative [308]. These results indicate the wave-particle duality, all matter exhibits both wave and particle properties. However, we notice the wave-particle duality mostly on the micro-scale.

As a consequence, the summation rule of classical probability theory is violated, resulting in one of the most fundamental laws of quantum mechanics, see [34]. Quantum physics by itself does not offer any justification or explanation beside the statement that it just works fine.

4.4.1 Schrödinger's Cat Paradox

Measurements always find the physical system to be in a definite state, they do something to the wave function. The best known example of this type kind of this "something" is the Schrödinger's cat paradox [243], it tells us that our universe is non deterministic. A Geiger counter measures the decay of a radioactive substance. There is a fifty percent chance that, in a given time frame, decay is measured. The Geiger counter is connected to a device that kills the cat, if decay is measured. Because the cat and the Geiger counter are in a closed room, we do not know whether the cat is dead or alive. Each of these possibilities is associated with a specific fifty percent probability.

The cat is dead and alive at the same time. How can this be? One can think that there is one world where the cat is alive and another where the cat is dead. They

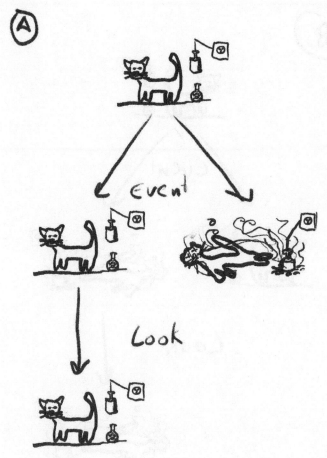

Figure 4.3 A Geiger counter measures the decay of a radioactive substance. There is a fifty percent chance that, in a given time frame, decay is measured. The Geiger counter is connected to a device that kills the cat, if decay is measured. At some point the decay is measured or not; this corresponds to an event that happens or not. Because the cat and the Geiger counter are in a closed room, we do not know whether the cat is dead or alive. Each of these possibilities is associated with a specific fifty percent probability. The cat is dead and alive at the SAME time. How can this be? One can think that there is one world where the cat is alive and another where the cat is dead. They are at the same time present in the closed room. The two states are "really" present at the same time; one says that the cat is in a superposition state. However, if we look into the room the cat is alive (A).

are at the same time present in the closed room. The two states are "really" present at the same time, one says that the cat is in a superposition state. A measurement always finds either an alive cat or a dead cat with a probability of fifty percent, see Figures 4.3 and 4.4.

A Laplace's demon cannot exist in quantum physics, the quantum universe is nondeterministic, one cannot determine if the cat is alive or dead. The measurement

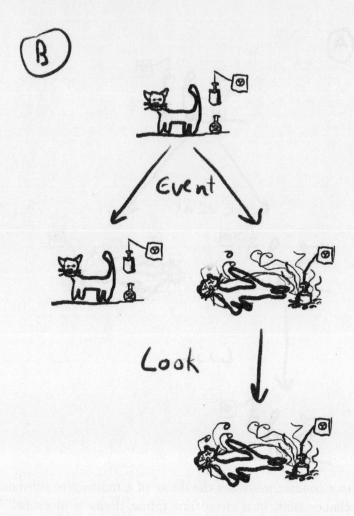

Figure 4.4 The cat is dead and alive at the SAME time; if we look into the room the cat is dead (B).

is nondeterministic, it represents a real random event without any causal effect. True randomness exists.

Indeed, the cat can represent the smallest information unit bit, 0 for dead and 1 for alive. In quantum computation, there are simpler methods to represent a bit than by a cat. A bit can be represented by a spin state of a particle, either spin down or spin up. In a closed room, we can map the spin in the two states, spin down and spin up, they are present at the same time. Such a bit is called a qubit. We can use these two states to perform a computation in parallel. When measured at the end of the computation, the spin is always in one of two possible states: Spin-up or spin-down. We can manipulate the probabilities in such way, that the results after the measurement become less random. We can also use several qubits to represent many states at the same time, eight qubits can represent 256 states at the same time, sixteen qubits 65,536 states. We could do a computation over 65,536 states at the same time. This sounds great, but there is a catch, the manipulations of the probabilities are not for free and slow down the computation in most cases.

4.4.2 Interpretations of Quantum Mechanics

During the measurement true randomness is present, after the measurement the physical system is always in a definite state. Measurement doses something to the wavefunction. This something is not explained by quantum theory itself, however there are two common interpretations:

- The most popular interpretation, the Copenhagen interpretation, claims that quantum mechanics is a mathematical tool that is used in the calculation of probabilities and has no physical existence; all other questions are metaphysical [132] and should be avoided.

- The many-worlds is less popular due to some philosophical difficulties. The many-worlds theory views reality as a many-branched tree in which every possible quantum outcome is realized [53, 96]. Every possible outcome to every event exists in its own world. In one world, randomness exists, but not in the universe (multiverse) that describes all possible worlds [83].

For the time being, we will follow the Copenhagen interpretation and use quantum mechanics as a mathematical model.

4.4.3 Observables

The measurement corresponds to the collapse of the state vector, it is not reversible, and it is not consistent with the unitary (linear) time evolution. A measurement is performed by observables. Observables are the things that are measured. They are machines with an input and output. The input is a state vector, and the output is number that indicates the measured state. Different observables can measure the input from different perspectives by rotating the coordinate system. Different perspectives can lead to different results [305].

If we measure a state by an observable, we gain information from that perspective. The measuring leads to collapse in one perspective and prohibits the full measurement from another perspective. This is the Heisenberg uncertainty principle; we cannot measure two perspectives at the same time. Different perspectives can represent different properties of a particle, like momentum and location of moving particles.

4.4.4 Hydrogen Atom

An hydrogen atom is composed of positively charged nucleus, the proton and an electron that is negatively charged. Atoms have no similarity to the solar system as lectured in the school, since gravity is negligible. Why does the electron not fall into proton? Atoms cannot exist according to classical physics [85].

An electron has more than one speed or more than one position due to the Heisenberg uncertainty principle. In a hydrogen atom, a proton is in the middle of a spreading cloud of instances of single electron. The cloud spreads due to its uncertainty-principle. The spreading stops when the uncertainty-principle diversity is balanced by its attraction on the proton.

4.4.5 Quantum Tunneling

This Heisenberg uncertainty means that some energy can be borrowed, to overcome some mountain and go out of a minimum as long as we repay it in the time interval [134]. To illustrate quantum tunneling, let us imagine a ball at the bottom of a hill. If we push the ball, it will begin to roll up the hill. If we push it hard enough, then the ball rolls over the hill. It means that the energy of the ball was greater than the potential energy needed to roll over the hill. If we do not push hard enough, then the ball cannot get over the hill since its energy is smaller than the potential energy needed to get over the hill. Instead of a ball, we will deal with an electron and instead of a hill we have a barrier represented by a repulsive electric field. If we give the electron enough momentum so that it would make it through the field, we see most of the electron goes through. However, some electrons do not make it and are reflected due to the time-energy uncertainty. If we give the electron not enough momentum so that it could not make it through the field, we see some of the electron has managed to make it through the barrier due to the time-energy uncertainty. Quantum tunneling allows classical energy barriers to be overcome to some extend by the time-energy uncertainty. A particle is a wave that can tunnel through a region in which the potential energy function exceeds its total energy. The wave function has some probability of tunneling through a barrier rather than over it. The quantum tunneling plays an important role in quantum machine learning and quantum biology.

There is a quantum annealer system called "D-Wave," which speeds up some optimization problems in machine learning [87,146,273]. D-Wave is not a general quantum computer, the system is based on quantum annealing, which finds the global minimum of a function [43,144]. Quantum annealing based on quantum tunneling attempts to avoid local minima using a quantum fluctuation parameter, which replaces a state by a randomly selected neighboring state.

Per-Olov Lowdin proposed 1963 a new field of study called quantum biology based on the idea of proton tunneling as a mechanism for DNA mutation. Electron tunneling in biological systems in proteins was first reported in [182]. Enzymes may use quantum tunneling to transfer electrons for long distances. In photosynthesis, electron tunneling plays an important role. In some enzymes, quantum effects may contribute to enzymatic catalysis.

4.5 QUANTUM COMPUTATION

Richard Feynman asked in the early eighties whether a quantum system can be simulated on an imaginary quantum computer. Today, 40 years later first small quantum computers begin to appear. A quantum computer represents information by qubits.

A qubit can be represented by the spin state of a particle, either spin down or spin up. Several qubits to represent different states at the same time. The qubits need to be coherent for a long enough time so that a computation can take place. During recent years, a huge progress was achieved in keeping the quits coherent, opening the road to a universal quantum computer.

4.5.1 Qubits

A qubit is represented abstractly by a two-dimensional vector. The first dimension of the vector corresponds to zero and the second to a one. The values of the two dimensions correspond to the probabilities that when measured, one of the two states is present. We can combine a qubit to several qubits by a tensor operation. Two qubits have four possible states, the first dimension represents the state zero-zero, the second zero-one, the third one-zero and the fourth one-one. They are described by a four-dimensional vector. Four qubits would represent sixteen different combinations, eight 256 combinations, represented by a 256-dimensional vector and so on. The number of possible states grows exponentially in relation to the number of qubits, with the number of states being equal to two power the number of represented qubits [305]. In a vector representing several possible states, each dimension corresponds to the probabilities of measuring a state. The probabilities values are represented by amplitudes. The relation between an amplitude and the probability is the absolute value of the amplitudes power two. For example, the probability value 0.25 corresponds to an amplitude 0.5, since $|0.5|^2$ is 0.25. The amplitude can be negative, since $|-0.5|^2$ is 0.25 or even imaginary numbers since $|-0.5 * i|^2 = 0.25$. As a consequence of this relation, the Euclidean length of the vector representing the probabilities is always one, all probabilities sum to one. If a dimension of the state vector represents the probability zero, the corresponding state will be never measured. If one dimension of the state vector represents the probability one, all other dimensions represent the probabilities zero, and the corresponding dimension will be measured for sure. If the represented probabilities are bigger than zero or smaller than one, they will be measured with certain probability. Repeating the same experiment several times the measured values converge to the represented probability values, the individual measurement is random. A qubit with the probability values 0.5 (amplitude $\sqrt{0.5}$) and 0.5 (amplitude $\sqrt{0.5}$) represents a true random process when measured. True random numbers are generated during the measurement.

4.5.2 Mixed States

When our knowledge of a system is incomplete, we represent a state by probabilistic mixture (combination) of different vectors. Such a state is called mixed and can be represented efficiently by a matrix [305]. In quantum computing, mixed states (opposite to quantum information) do not play any essential role. In this context, a state that is represented by a vector is called a pure state.

4.5.3 Linear Operators

The computation on the coherent qubits (before the measurement) is described by linear operators that change the distribution of the quantum probabilities represented by amplitudes in a linear way [305]. The operators are represented by orthogonal matrices for real amplitudes (inverse of the matrix is the transpose of the matrix), for complex amplitudes by unitary matrices (inverse matrix is the conjugate transpose of the matrix).

A linear operation that would produce a copy of an arbitrary quantum state is not possible. We cannot copy an unknown amplitude distribution of a state. This has profound implications in the field of quantum computing, since we cannot reuse an arbitrary quantum state.

The operators can be decomposed into smaller operators acting on some parts of the state vector representing a group of qubits. Such matrices operate on one or a group of qubits and are represented by quantum gates. Several qubits are abstractly described by a register. If we can decompose a register into a tensor product of several qubits, then the quantum gates can operate in parallel on these qubits. A computational step corresponds to such an operation, it can consider exponentially many states at the same time. After the computation, an additional algorithm changes the amplitudes in such a way that the solution of the computation can be measured. The additional algorithm is usually based on one of two principles that will be discussed later.

4.5.4 Entanglement

It can happen that we cannot decompose a register into individual qubits after certain linear operations. For example, a register of two qubit is decomposable if it can be represented as a tensor product of two qubits. A state that is not decomposable is called entangled. If two qubits are entangled in a state, then observing one of them will result in the same value of the other one. Both qubits behave as one unit and are called an *ebit*. The two qubits in an *ebit* behave as one unit, even if the qubits are separated. Once either qubit of an ebit is measured, the states of both qubits become definite. Experiments have shown that this correlation can remain even if the qubits are separated over a distance of several kilometers. It is possible to teleport a qubit from one location to another using an *ebit* [30]. More than two qubits can be entangled.

Quantum collapse during measurement is a non-local force. A non-local interaction is not limited by the speed of light, and its strength is not mediated with distance. This arrangement conflicts with Einstein's theory of special relativity, which states that nothing can travel faster than light. The conflict is resolved by the fact that one cannot use an ebit to send any information. If two qubits of an ebit are separated over a distance in two places, A and B, and there are no other means of communication, then measuring the qubit on place A determines the outcome on place B, but at place B, the outcome is unknown. Measuring at place B is a random process without the knowledge of the results of place A. Also to preform teleportation we need a conventional channel to send an information how to map the teleported qubit in a correct state.

4.6 TWO PRINCIPLES OF QUANTUM COMPUTATION

In quantum computation, there are two principles (algorithms) that speed up the computation by changing the probability distribution so that we can measure the desired solution. In addition to their being "faster" and their ability to generate true randomness, quantum computers have identical computational power to a Turing machine. The two principles are the Quantum Fourier Transform and Grover's amplification algorithm [305].

4.6.1 Quantum Fourier Transform

Quantum Fourier transform can be used to determine the period of a periodic function in polynomial time, see Figure 4.5. This exponentially faster than a conventional computer! It is used in one of the most famous machine learning algorithms that speeds up the solution of linear equations.

It also the framework for the factorization algorithm on which the famous Shor's algorithm is based [253, 254]. Shor's algorithm breaks conventional cryptographic codes efficiently. Such codes cannot be broken by conventional computers since the calculations would require an exponential amount of time.

A periodic function can be represented in the frequency space. The frequency is the number of occurrences of a repeating event per one unit of time. If something

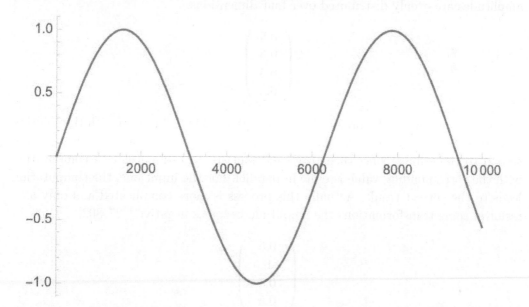

Figure 4.5 On a conventional computer we need to compute around 10,000 steps to determine the period of an unknown periodic function. On a quantum computer we need only 9 to 10 steps (log 10,000).

changes rapidly, then we say that it has a high frequency. If it does not change rapidly, i.e., it changes smoothly, we say that it has a low frequency. The discrete Fourier transform converts discrete time-based or space-based data into the frequency domain.

A periodic function can be represented as a superposition of qubits and their values of amplitudes (representing the probabilities). The quantum Fourier transform is represented by a unitary matrix that describes the discrete Fourier transform. The matrix can be efficiently decomposed so that the computation can be done in parallel. During this computation, the amplitudes that represent the periodic function are converted into amplitudes that represent the frequency of the function. For periodic function, the amplitudes representing the frequency of the function have positive value, all other amplitudes are zero. By measuring the register with high amplitude values we can reconstruct the period as well as the phase (Kitaev's phase estimation algorithm [147, 150]).

4.6.2 Grover's Amplification Algorithm

The second principle is described by the Grover's amplification algorithm [117–120]. We represent n state by a superposition, each state has the same real positive amplitude. A parallel computation is applied to all the n states, and the state with the solution is marked by a minus sign, the amplitude is now negative. We then apply a linear operation that is based on Householder reflection, by the reflection the value of the marked amplitude grows linearly.

Let us imagine a toy example with $n = 4$. Before we start the computation, the amplitudes are evenly distributed over four dimensions,

$$\begin{pmatrix} 0.5 \\ 0.5 \\ 0.5 \\ 0.5 \end{pmatrix}$$

each corresponding to the probability $0.25 = |0.5|^2$ [303]. The state with the solution is marked by a negative sign. The first dimension represents the number zero, the second to the number one, then three and four. In each dimension, a computation with the corresponding value is done in parallel. For the input one, the computation leads to the correct result. Actually this process is more complicated and only as a result of some transformations the amplitude becomes negative [227, 302].

$$\begin{pmatrix} 0.5 \\ -0.5 \\ 0.5 \\ 0.5 \end{pmatrix}$$

The Grover's iterate that is based on Householder reflection described by an unitary

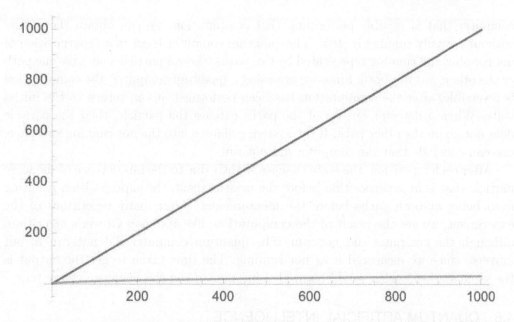

Figure 4.6 For n possible values we need n steps to find a solution on a conventional computer (upper line). On a quantum computer, we need only $\sqrt{n}$ steps (small and nearly unnoticeable curve below the line).

matrix amplifies the amplitude into

$$\begin{pmatrix} 0 \\ 1 \\ 0 \\ 0 \end{pmatrix}$$

and a measurement indicates the solution in the corresponding dimension. For dimensions higher than four ($4 \ll n$), the operations of marking and Householder reflection must be repeated $\sqrt{n}$ times, since at each step the amplitude only grows linearly. After $\sqrt{n}$ steps we measure the solution, see Figure 4.6.

Grover's amplification algorithm is optimal, one can prove that a better algorithm cannot exist [29,39]. It follows that using a quantum computer $NP-complete$ problems remain $NP-complete$. This is also due to the fact that it assumed that the factorization algorithm of the previous section is not $NP-complete$.

The algorithm guarantees us a quadratic speed up over a classical computer that would require n steps. If several solutions exist, we can determine their number k efficiently with the quantum Fourier transform, then apply Grover's amplification algorithm with the reduced costs $\sqrt{n/k}$ to measure just ONE solution out of several possible solutions.

4.7 COUNTERFACTUAL QUANTUM COMPUTATION

It's possible to build a quantum computer that computes without running. This approach is described by counterfactual quantum computation. If we have a quantum computer that is capable performing that computation, we can obtain the results without actually running it [195]. The quantum computer is set in a superposition of not running and running represented by two paths where a particle may take one path or the other, but not both. Since we are using a quantum computer, the computation is reversible, after the computation has been performed, it can return to the initial state. When a detector on one of the paths catches the particle, then the particle does not go on the other path. If the system collapses into the not running subspace, one can conclude that the computer did not run.

After each repetition, the state changes slightly due to the interference of the wave particle that is in superposition before the measurement, the superposition resulting from being at both paths before the measurement. After many repetitions of the experiment, we get the result of the computation, like applying Grover's algorithm, although the computer did not run. The quantum computer did not run in our universe since we measured it as not running. The time taken to get the output is the same as that, which would be needed in order to run the quantum computer.

4.8   QUANTUM ARTIFICIAL INTELLIGENCE

In the future, quantum artificial intelligence will offer us new possibilities. However, these possibilities are not as ground-breaking as it is sometimes believed. Mainly the applications of quantum artificial intelligence will be faster. The quantum computers will be faster because they can simultaneously encode many inputs of a problem and perform the calculation on all inputs in parallel. However, it is not possible to access the correct solution without additional costs. The costs are related to the number of simultaneous problems that are solved.

4.8.1   Quantum Machine Learning

Using Grover's algorithm, we can achieve a quadratic speed-up.

If a collection of n vectors is represented in a superposition, we can determine for a query vector the most similar vector with a quadratic speed up. The naïve assumption that the speed-up is quadratic is not realistic.

Most quantum machine learning algorithms are based on Grover's algorithm suffer from the input destruction problem (ID problem) [1, 3, 316]:

- The input (reading) problem: The amplitude distribution of a quantum state is initialized by reading n data points. Although the existing quantum algorithm requires only $\sqrt{n}$ steps and is faster than the classical algorithms, n data points must be read. Hence, the complexity of the algorithm does not decrease.

- The destruction problem: We are required to read n data points and are allowed to query only once because of the collapse during the measurement (destruction).

By ignoring the input problem, which is the main bottleneck for data encoding, theoretical speed ups can be analyzed [244]. For example, we can assume that the data is already present or was generated by a quantum computer. With quantum data, we can use Grover's algorithm for quadratic speed up over classical machine learning algorithms [316].

Some quantum machine learning algorithms overcome these problems by the efficient preparation of data as described in the next section.

4.8.1.1 Linear Algebra-Based Quantum Machine Learning

Linear algebra-based quantum machine learning can exhibit theoretical exponential speed ups over the classical counterparts. It solves linear sparse equations efficiently. Sparse equations are required to overcome the ID problem.

The algorithm is one of the main fundamental algorithms expected to provide a speedup over their classical counterparts. In the honor of its inventors Aram Harrow, Avinatan Hassidim, and Seth Lloyd, it is called the HHL algorithm [129]. HHL algorithm is one of the most useful subroutines for any quantum machine learning algorithm because almost all machine learning algorithms involve matrix multiplication or some form of optimization in which a matrix is inverted.

The algorithm has already been implemented in existing quantum computers/devices [209].

A related algorithm is the quantum PCA. Both algorithms are based on extended quantum Fourier transform, (Kitaev's phase estimation algorithm).

A quantum computer can as well estimate efficiently a scalar product between two vectors in high dimensional space. The vectors are the result of a certain mapping into in a huge dimensional space during the computation. It is impossible to estimate the scalar product between such huge vectors on a classical computer due to memory limitations. On a quantum computer, we get an exponential quantum advantage in evaluating such a scalar product [244].

4.8.1.2 Quantum Coprocessor

In the future, the quantum basic linear algebra subroutines will be represented by a quantum coprocessor for extensive and nontraceable computation routines in machine learning. Such "mathematical" quantum coprocessor can be used with a conventional computer [305].

4.8.2 Symbolical Quantum Artificial Intelligence

Most algorithms in symbolical AI are based on search, we have a set of possible rules that we can apply. We want to reach a goal by applying the correct rules. This corresponds also to the cognitive models of human thinking that are described by the production systems [305]. Imagine a robot in a labyrinth with one way out. At each point, it has two possible rules it can apply, either it goes to left or too right. In the worst case, the robot does the search over all combinations of rules to find a way out of the labyrinth. If we have two rules and we want to search to a depth of three, we have at the first level two possibilities. Then we have again two possibilities. It means that at the depth of two we have two plus four possibilities. At the depth, three six plus eight possibilities. This kind of search is described by an algorithmic structure

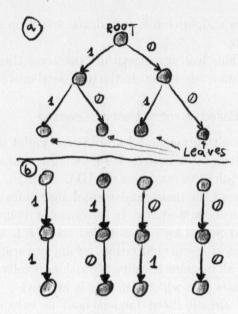

Figure 4.7 Search tree for $B = 2$ and $m = 2$. Each question can be represented by a bit. Each binary number (11, 10, 01, 00) represents a path from the root to the leaf.

called a tree, see Figure 4.7. If humans model a problem, they cannot consciously search very deep because of the limitations the short-term memory. The choice can be based intuition described by a heuristic function, however it is very difficult to invent a heuristic function. Instead a quantum computer could speed up the search without any heuristic function.

4.8.2.1 *Tree Search and the Path Descriptors*

Nodes and edges represent a search tree. Each node represents a state, and each edge represents a transition from one state to the following state. The initial state defines the root of the tree. From each state, either B states can be reached, or the state is a leaf. From a leaf no other state can be reached [305]. B represents the branching factor of the node, the number of possible choices. A leaf represents either the goal of the computation or an impasse when there is no valid transition to a succeeding state. Every node besides the root has a unique node from which it was reached, which is called the parent. Each node and its parent are connected by an edge. Each parent has B children. If $B = 2$, each of m questions has a reply of either "yes" or "no" and can be represented by a bit (see Figure 4.7). The m answers are represented by a binary number of length m. There are $n = 2^m = B^m$ possible binary numbers of length m. Each binary number represents a path from the root to a leaf. For each goal, a certain binary number indicates the solution. For a constant branching factor $B > 2$, each question has B possible answers. The m answers can be represented by a base-B number with m digits. For example, with $B = 8$, the number is represented

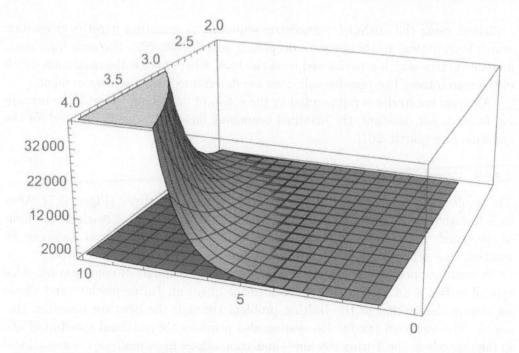

Figure 4.8 For branching factor B from 2 to 4 and depth of the tree search m from 1 to 10. The cost on a conventional computer are $n = B^m$ (upper plane). On a quantum computer, we need only $\sqrt{n} = B^{\frac{m}{2}}$ steps (plane below).

in an octal numeral system. These numbers represent all paths from the root to the leaves.

4.8.2.2 Quantum Tree Search

In a quantum computation, we can simultaneously represent all possible path descriptors. There is one path descriptor for each leaf off the tree. Using Grover's algorithm, we search through all possible paths and verify whether each path leads to the goal state. This type of procedure is called a quantum tree search [267]. For $n = B^m$ possible paths, the costs are (approximately) $\sqrt{n} = B^{\frac{m}{2}}$, see Figure 4.8.

A constraint of this approach is that we must know the depth m off the search tree in advance. The constraint can be overcome by iterative deepening. In an iterative deepening search, we gradually increase the limit of the search from one to two to three to four and continue to do so until a goal is found. For each limit, a search is performed from the root to the maximum depth of the search tree. If the search is not successful, a new search is initiated with a bigger depth.

During the iterative deepening search, the states are generated multiple times [157, 235]. The time complexity of the iterative deepening search is of the same order of magnitude as a search to the maximum depth [157], as explained by Richard E. Korf: "*Since the number of nodes on a given level of the tree grows exponentially with depth, almost all time is spent in the deepest level, even though shallower levels are generated an arithmetically increasing number of times.*" The paradox can be explained using the arithmetic-geometric sequence. A quantum iterative deepening search is equivalent to the iterative deepening search [268, 269]. For each limit max, a quantum tree search is performed from the root, where max is the maximum depth of the search tree. The possible solutions are determined using a measurement.

A second constraint is represented by the constant branching factor. If the branching factor is not constant, the maximal branching factor B_{max} must be used for the quantum tree search [267].

4.8.2.3 *Complete Model of Computation*

The result of the computation is represented by a path descriptor (Figure 4.7). After each iterative step, Grover's algorithm is performed. If no goal is reached, the limit of the search is increased, and the procedure is repeated. If the goal state can be reached, the corresponding path descriptor is accessed by Grover's algorithm.

A quantum production system represents a general model of computation. This type of system is an alternative approach to the quantum Turing machine and allows an elegant description of the Halting problem through the iterative quantum tree search. The quantum production system also provides the maximal speedup of $\sqrt{n}$ in the case where the Turing machine simulation allows for n multiple computational branches [269].

4.8.2.4 *Cognitive Models of Human Thinking*

SOAR describes the architecture of the mind. An extension of the SOAR cognitive architecture by a quantum tree search would lead to a hybrid architecture. The iterative quantum tree search would be invoked if an impasse is present. An impasse corresponds to the lack of applicable knowledge, it cannot be decided which rule to use. The iterative quantum tree search would model the unconscious mind as described by Poincaré. Such a hybrid approach would speed up the learning process without a need for domain-specific control knowledge. Computationalism states, that we can simulate the mind by a Turing machine. We extend computationalism by the idea that we can simulate the mind by a quantum Turing machine. In this extension, we would model the consciousness and unconsciousness part of our mind [305].

Most neuroscientists believe that the brain is too warm and wet for any quantum algorithmic process like the quantum tree search. The necessary quantum coherence seems only possible in highly protected and frigid environment.

4.8.3 Uncertainty in Artificial Intelligence

The wave function in quantum mechanics represents a superposition of states. If unobserved it evolves smoothly and continuously; however, during the measurement, it collapses into a definite state. According to most physical textbooks, the existence of the wave function and its collapse is only present in the microscopic world and is not present in the macroscopic world. More physical experiments indicate that wave functions are present in the macroscopic world [284]. Physical experiments state that the size does not matter and that a very large number of atoms can be entangled [6, 108].

Clues from psychology indicate that human cognition is not only based on traditional probability theory as explained by Kolmogorov's axioms but additionally on quantum probability (amplitudes), see [48–52, 149]. For example, humans when making decisions violate the law of total probability. The violation can be explained as a quantum interference resulting from the quantum probabilities (amplitudes) [51].

A unified explanation of human interference using quantum probability and classical theories was proposed for the first time in [279, 308]. The hierarchy of mental representations, from quantum-like to classical, is built and allows the combination of both Bayesian and non-Bayesian influences. In this model, classical representations provide a better account of data as individuals gain familiarity with it. There is a distinction between unknown, which is considered as the truth value, and ignorance, which is considered as the lack of knowledge or as being unaware. An event can be true false or unknown.

- For unknown events, the law of total probabilities is applied.

- For the events of which we are unaware (ignorance), we apply quantum-like models with the interference resulting from the quantum probabilities (amplitudes). We determine the possible value of the wave that produce the interference by the law of maximum uncertainty of quantum-like systems.

An unknown event is not known to us because we do not have enough information. Ignorance means that we cannot achieve this information. Ignorance is not a truth value at all. An ignorant event means that we do not know the value at all. The relation is equivalent as the relation between pseudo randomness and true randomness. Pseudo randomness appears to us as random due to our lack of information. True quantum randomness corresponds to ignorance. The ignorance corresponds to prediction in the future. The emerging field that studies the quantum-like models that describe the human interference is called quantum cognition. Quantum cognition leads to the conclusion that a wave function can be present at the macro scale of our daily life. Quantum-like models indicate how to describe probabilistic reasoning in artificial intelligence.

4.8.4 Quantum Robots

The term "robot" was used for the first time for machines capable of carrying out a complex series of actions automatically 1920 play R.U.R. by the Czech writer, Karel Čapek, from Slavic "robota," which means work. A quantum robot is a mobile quantum system that includes a quantum computer to speed up the computation [25]. There are many different types of robots, a humanoid robot resembles the human body in shape. Such a robot can ease the interaction with humans, quantum cognition would make its behavior even more human. Another interesting type of a robot is a space probe, it is a robotic spacecraft that can explore the further space, like different planets or even the interstellar space. A quantum robot would use a mobile quantum computer and quantum coprocessor that would dramatically increase its capacity. The quantum coprocessors would improve its mathematical computation, allow faster computation of trajectories including complicated kinematics, and so allow faster planning of movements. Quantum SOAR architecture described by a quantum computer would allow faster planning of actions, for example, flying a plane or support better face-to-face dialogues with humans. Quantum robots will be faster, more accurate, and would be able to de multitask better than the standard robot.

Such a quantum robot would be far superior to us humans when solving certain tasks, since it seems that human brain does not implement any quantum algorithmic process, it is simply too warm and wet. Also, the neurons in human brain do not perform a reversible computation, neither do neural networks.

4.9 QUANTUM BRAIN

The initial motivation for relating quantum theory to consciousness was related to free free will in a deterministic world since the quantum world is nondeterministic. Quantum events are random; however, their indeterminism is constrained by statistical laws. Could quantum randomness open novel possibilities for free will? Danko Georgiev proposes that interaction between the central nervous system and the surrounding environment performs a quantum measurement upon the neural constituents [107]. According to Danko Georgiev alternative physical outcomes provide varying amounts of free will that a quantified by the expected information gain. However, this kind of randomness is problematic for goal driven behavior.

4.9.1 Dualism and Quantum Physics

The dualism of mind and matter involves the unresolved problem of how the mind and brain could interact. The relationship of union between the mind and the body of a person is resolved by with physical state reductions during the measurement of a quantum event. London and Bauer [174] proposed 1939 that it is indeed human consciousness that completes the quantum measurement process.

4.9.1.1 Stapp

Henry Stapp assumes that the conscious intentions of a human being can influence the activities of his brain [260]. The measurement of quantum events in the brain is not random but positively or negatively biased by the positive or negative values in the minds of the observers. This hypothesis leads into mental influences on quantum physical processes. Free will corresponds to an intentional decision for an action by the mind preceding the action itself.

4.9.1.2 Beck and Eccles

Sir John Eccles proposed of how there can be mind-brain action without violating the principle of the conservation of energy [91]. The interaction takes place at the information transfer at the synaptic cleft of chemical synapses of a neuron. In order to propagate a signal, a chemical transmitter (glutamate) is released at the presynaptic terminal, the release process is called exocytosis [24,91].

This mechanism can be described in a statistical way by thermodynamics or quantum mechanics. In the proposed model, quantum processes are relevant for the release of the chemical transmitter (exocytosis) that are related to states of consciousness. Eccles assumes the consequent voluntary actions of consciousness manifest itself by increases of the probability of vesicular emission in the thousands of synapses on each pyramidal cell. The consciousness controls the release of the chemical transmitter, it interacts with the brain. Eccles assumes that contemporary physics cannot detect, measure or predict the supposed mental forces of the consciousness. Consciousness manifests itself in mental intentions by interacting with the matter. Eccles suggest that there is a two-way traffic between mind and the matter-energy system.

Jahn and Dunne [142] claim that they have created statistically significant results that suggest a direct causal relationship between subjects intention and otherwise random results. The parapsychological experiments were based on random event generators. A comprehensive review of psychic phenomena from an engineering perspective was published [142]. However, till today the results could be not reproduced in other laboratories and several flaws were detected.

4.9.2 Non-Computable Aspects of New Physics

Sir Roger Penrose argues that elementary acts of consciousness cannot be described by a Turing machine [212,213]. Such noncountable elementary acts of consciousness are only present in the wave function collapse (measurement). Penrose assumes that the wave function collapse is not random, it represents an objective reduction. The wave function describing the state of a quantum system is approximate. It works well for microscopic systems, but progressively loses its validity when the mass and complexity of the system increases. Penrose assumes that gravity is responsible for the collapse of the wave function and that during the collapse unobserved part is removed. A spatial superposition creates the superposition of two different spacetime curvatures that are spontaneously collapsed by gravity. The gravitation-induced reductions of coherent superposition states are present in microtubuli of the neurons. Microtubules are a substance or material consisting of very large molecules of tubulin (tubulin protein superfamily of globular proteins) and provide structure and shape to the neurons. Microtubuli ensure that quantum states remain long enough to become reduced by gravitational influence rather than by interactions with the environment within the brain. They were suggested as a possible place by Hameroff and Penrose [125]. The respective quantum states are assumed to be coherent superpositions of tubulin states, ultimately extending over many neurons. Their simultaneous gravitation-induced collapse is interpreted as an individual elementary act of consciousness. Quantum coherence in microtubules and its gravitation-induced collapse are related to the consciousness, the description of the process is based on new undiscovered laws of physic. An element of proto-consciousness takes place whenever a decision is made in the universe. The model relies on the undiscovered laws of physic and is related to the traditional dualism.

CHAPTER 5

Multiverse

T HE modern version of the wave-function collapse in quantum mechanics is based on decoherence and leads to the multiverse interpretation of quantum mechanics.

5.1 EVERETT MANY-WORLDS

Measurements always find the physical system to be in a definite state, they do something to the wave function. Let us remember the Schrödinger's cat paradox, the cat is dead and alive at the same time. However, a measurement always finds either an alive cat or a dead cat with a probability of fifty percent.

Let us imagine instead of the cat Andreas in the closed room. A Geiger counter inside the room measures the decay of a radioactive substance. There is a fifty percent chance that in each time frame decay is measured. The Geiger counter is connected to a device that handles Andreas a bottle of whisky if decay is measured. Because Andreas and the Geiger counter are in a closed room, we do not know whether Andreas is drunk or sober. Each of these possibilities is associated with a specific fifty percent probability. And of course, Andreas is bored in the closed room, likes whisky, and drinks the whole bottle.

A measurement always finds either Andreas drunk or sober with a probability of fifty percent. Who measures the state by entering in the closed room, it is Manuela, Andreas wife. Let us imagine Manuela finds Andreas drunk. This has serious consequences for the drunk Andreas, since Manuela does not like drunk people. Manuela is angry.

But what happened with the sober Andreas? He mysteriously disappeared, he was there, and then he is gone as if he never existed. The Copenhagen interpretation, claims that quantum mechanics is a mathematical tool that is used in the calculation of probabilities and has no physical existence; all other questions are metaphysical [132] and should be avoided. A mystery happened. Science does not offer us any explanation.

Let us change the perspective and see what Andreas does. Andreas is in a closed room, there is a strange apparatus in the room. Andreas is bored. Suddenly a door of the apparatus opens, and a bottle of whisky is presented to him. He opens the bottle, drinks a gulp, then another one. The life was never more beautiful for him than in this moment, but then the disaster strikes. The door of the room opens, and

DOI: 10.1201/9781003244547-5

his wife Manuela enters, and she is not happy seeing her husband drunk. We do not know how the history ends, but certainly this history involves a drunk Andreas and an angry Manuela.

Let us redo the story. Andreas is in a closed room, there is a strange apparatus in the room. Andreas is bored. Time passes very slowly, nothing happens, like in a waiting room that is an image of the hell, but then the happy moment arrives. The door of the room opens, and his wife Manuela enters, and she is happy seeing her husband. We do not know how the history ends, but certainly this history involves a happy Andreas and Manuela. So in the end there are two histories.

For the Copenhagen interpretation to be true, we have to replace one of the two stories by another one. Imagine Manuela finds her husband drunk. We know the story about the drunk Andreas. But what happened to the sober Andreas? Let us see. Andreas is in a closed room, there is a strange apparatus in the room. Andreas is bored. Time passes very slowly, nothing happens, like in a waiting room that is an image of hell, but then the whole world disappears in nothingness. What happened? Which explanation seems more plausible, the magical disappearance or the division of histories?

5.1.1 Schrödinger's Cat and the Multiverse

Back to our poor the Schrödinger's cat, we can represent the states using simple mathematical addition and multiplication. The state of the cat in superposition is represented as $Cat_{dead} + Cat_{alive}$ after measuring by the observer we get the following interpretation, either

$$(Cat_{dead} + Cat_{alive}) \cdot observer = Cat_{dead} \cdot observer$$

means observer is either with the dead cat or

$$(Cat_{dead} + Cat_{alive}) \cdot observer = Cat_{alive} \cdot observer$$

means observer is with the cat that is alive according to the Copenhagen interpretation. Every little child knows that both equations are wrong.

The problem arises by the description of observer-independent world, like there is some microscopic quantum world and the macroscopic world is not part of it. We have to consider the entire system, including the measuring device (observer), as a single quantum system [37]. It should be instead

$$(Cat_{dead} + Cat_{alive}) \cdot observer = Cat_{dead} \cdot observer + Cat_{alive} \cdot observer$$

It leads to the many-worlds theory (multiverse) [96], [53].

5.1.2 Hugh Everett III

Because the predictions of quantum mechanics have been compared with experiments many times and have always been found to be valid, most physics do not really care about the interpretation problem of the measurement and follow the Copenhagen

Figure 5.1 Hugh Everett III (1930–1982), one of the greatest physicsists of the 20th century, suggested the many-worlds theory (multiverse) or many-worlds interpretation (MWI).

interpretation. They use the mathematical formulas without any thoughts about the unexplained measuring process.

Hugh Everett III (1930–1982), was a graduate student working with Prof. John A. Wheeler at Princeton University in the mid 1950s, see Figure 5.1. For Everett the measuring paradox was a "magic" process that tied to the role of the observer. Everett did not make the distinction between the microscopic and macroscopic worlds [53,96]. If quantum mechanics was true on the microscopic level in a laboratory, it should be also true on the macroscopic level of the whole universe. However, in a universe there is no external observer outside the universe. Everett assumed that there is no observer-caused collapse during the measurement and we have to consider all quantum states to be always noncollapsed. When an observer in the universe makes a measurement of the electron, he does not cause a collapse. Instead, he becomes correlated with the electron. Instead of an electron there is the electron and the observer. The observer is an integral part of the measured system but does not play any role during the measurement itself. All elements of a superposition are actual, none anymore "real" than another. After an observation no element of the final superposition is mysteriously selected to be awarded and the others condemned to oblivion. Everett broke with the Copenhagen interpretation and with the discontinuity of a wave-function collapse.

The observer bifurcates at each interaction with the superposed object, it is like a tree that branches and re-branches. Each of these new branch universes has a different history. The observer follows one path through the diverging branches and never perceive the splitting. Instead, he interprets the resolution into one particular outcome as a Copenhagen-style collapse. Each branch once formed with its own copy of the observer does not influence other branches. Each branch embarks on a different future, independently of the others.

For Everett, the argument that the theory is contradicted by experience as we are unaware of any branching process, is like the criticism of the Copernican theory that the mobility of the earth as a real physical fact is incompatible with the common sense interpretation of nature because we feel no such motion.

The suggested theory is called the many-worlds theory (multiverse) or many-worlds interpretation (MWI) [96]. According to the theory there should be a universe in which every physically possible event has happened.

In April 1956 a copy of the long version of Hugh Everett's thesis was sent to the Institute for Theoretical Physics in Copenhagen [22]. Alexander Stern was a physicist and engineer who spent many years during the 1950s and 1960s at Bohr's institute for Theoretical Physics in Copenhagen. He held a seminar on Everett's work, during which he recorded some of the reactions of members of the Institute to Everett's ideas. After the seminar, Stern wrote a long letter to Wheeler describing his own and his colleagues' dissatisfaction with Everett's theory. He dismissed Everett's theory as "theology" [22]. Some extracts of the letter [22]: "*In my opinion, there are some notions of Everett's that seem to lack meaningful content, as, for example, his universal wave function. ... his lack of an adequate understanding of the measuring process seem to appreciate the FUNDATALLY irreversible character and the FINALITY of a macroscopic measurement. ... The observer in quantum theory prepares the initial state of the system... If Everett's universal wave equation demands a universal observer, an idealized observer, then this becomes a matter of theology.*"

A short article almost identical to the final version of his thesis, was published in Reviews of Modern Physics [96]. It was a compromise between Everett and his supervisor Wheeler. After the paper appeared, it slipped into instant obscurity. Wheeler gradually distanced himself from association with Everett's theory. Everett received his Ph.D. in physics from Princeton in 1957 after completing his doctoral dissertation titled "On the foundations of quantum mechanics." Discouraged by the scorn of other physicists for the many-worlds interpretation (MWI), Everett ended his physics career after completing his PhD.

During March and April 1959, at Wheeler's request, Everett visited Copenhagen, in order to meet with Niels Bohr, the "father of the Copenhagen interpretation of quantum mechanics" [22]. The visit was a complete disaster; Everett many-worlds interpretation (MWI) was simply heresy to Bohr and the others at Copenhagen. Everett was described by Leon Rosenfeld as *stupid and not understanding the simplest things in quantum mechanics.*

In 1977, Everett was invited to give a talk at a conference Wheeler had organized at Wheeler's new location at the University of Texas at Austin [22],

Everett's talk was quite well received and influenced a number of physicists in the audience, including Wheeler's graduate student, David Deutsch, who later promoted the many-worlds interpretation to a wider audience. Everett, enjoyed the presentation; it was the first time for years he had talked about his quantum work in public. David Deutsch, the founder of the field of quantum computation (itself inspired by Everett's theory), states: *"Everett was before his time. He represents the refusal to relinquish objective explanation. A great deal of harm was done to progress in both physics and philosophy by the abdication of the original purpose of those fields: to explain the world. We got irretrievably bogged down in formalisms, and things were regarded as progress, which are not explanatory, and the vacuum was filled by mysticism and religion and every kind of rubbish. Everett is important because he stood out against it."*

The MWI received more credibility with the discovery of quantum decoherence in the 1970s and has received increased attention in recent decades, becoming one of the mainstream interpretations of quantum mechanics.

5.1.3 New Interpretation of the World

The explanation is simple, too simple? It leads to some philosophical questions that, for some scientists, create an uncomfortable feeling. We have the choice, between a metaphysical explanation or the multiverse explanation of different histories. The multiverse explanation removes the paradox of the measurement. The paradox is unscientific and leads to many metaphysical explanations, that the mathematical logic is wrong, the idea that quantum physics is not correct and incomplete. Multiverse theories have been widely criticized with the argument that current theories of particle physics are based on a version of quantum mechanics, which has a reputation for not being understandable, but what if we take quantum physics seriously? Following Occam's razor that the simplest explanation is usually the best one, we conclude that quantum physic is correct without the need of extending it. Humans do not like changes, for example, a new radical interpretation of the world. This behavior is not new.

5.1.3.1 Pythagoreans

Pythagoras of Samos (569–500 BC) was a mathematician and the founder of the Pythagoreans. Pythagoreans focused on mathematics, but also had some religious and mystic assumptions. They believed that the numbers rule the universe and that each number held great meaning, odd numbers were male, and even numbers were female. The Pythagorean theorem was used by the Babylonians and Chinese 1000 years before Pythagoras' time. However, it is assumed that Pythagoras was the first to offer a proof. He noticed, that if he takes a triangle with two legs off lengths 1 he derives $\sqrt{2}$ that is not a rational number. Pythagoreans believed that numbers were rational, but $\sqrt{2}$ is irrational. Since they worshiped the numbers, they treated this as a secret [190]. As the legend goes, Hippasus of Metapontum was murdered for revealing this secret.

5.1.3.2 Geocentric and Heliocentric Model

Claudius Ptolemy (100–170 AD) was a mathematician, astronomer as mentioned before in the section about algorithms. He published is a cosmological work with the title "Hypotheses of the Planets" dealing with the structure of the universe and the laws that govern celestial motion [196]. He presented a physical realization of the universe as a set of nested spheres, in which he used the epicycles of his planetary model to compute the dimensions of the universe. In his planetary model of the universe the earth is at the center of the universe. Under this geocentric model, the sun, moon, stars, and planets all orbit earth. The geocentric model was the predominant description of the cosmos in many ancient civilizations.

The notion that the Earth revolves around the Sun had been proposed as early as the third century BC by Aristarchus of Samos (310–230 BC). The heliocentric model opposed the geocentric model that placed the Earth at the center [198]. Despite the fact the geocentric model remained valid in Europe till sixteen century. It was supported by the catholic church; the humans were at the center of the world.

During the Polish Renaissance, the polish mathematician, astronomer, and Catholic Nicolaus Copernicus (1473–1543) wrote the book "Revolutions of the Heavenly Spheres" [65]. The book was first printed in 1543 in Nünemberg after his death. The book presented a mathematical model of a heliocentric system, an alternative model of the universe to Ptolemy's geocentric system, which had been widely accepted since ancient times, leading to the Copernican Revolution.

Later Johannes Kepler introduced elliptical orbits, and Galileo Galilei presented supporting observations made using a telescope. In February 1616, the Inquisition condemned heliocentrism as foolish and absurd in philosophy, and formally heretical since it explicitly contradicts in many places to the sense of Holy Scripture. Later in March 1616 the Pope banned all books and letters advocating the Copernican system, which he called "the false Pythagorean doctrine, altogether contrary to Holy Scripture." Robert Bellarmine (1542–1621) summoned Galileo, notified him of a forthcoming condemning the Copernican doctrine of the mobility of the Earth and the immobility of the Sun, and ordered him to abandon it and Galileo agreed to do so [99]. Bellarmine argued that he did not support the heliocentric model for the lack of evidence of the time, however he saw no problem if it was treated as a purely hypothetical calculating device and not as a physically real phenomenon. This statement reminds us of the Copenhagen interpretation.

The heliocentric model wastes a lot of resources, why the enormous waste of empty space, why not the simple economical world model as described in the book "Hypotheses of the Planets" by Claudius Ptolemy. The "Revolutions of the Heavenly Spheres" by Nicolaus Copernicus remained forbidden until 1758, the copy of the original manuscript is kept at the Jagiellonian University Library in Kraków.

Figure 5.2 David Deutsch is the founder of the field of quantum computation.

5.1.4 David Deutsch Argument

After Bohr, the Copenhagen interpretation "shut up and calculate" philosophy took over physics for decades. To delve into quantum mechanics as if its equations told the story of reality itself was considered sadly misguided. Quantum computers could be invented in nineteen-thirties, they were not due to the Copenhagen interpretation. David Deutsch, the founder of the field of quantum computation itself inspired himself by Everett's theory, see Figure 5.2. The quantum algorithms speed up comes from outsourcing of work, calculations take place in other universes. Entangled particles communicate among different universes, sharing information and gathering the results. A number of physics journals rejected some of Deutsch's early quantum-computing work, saying it was "too philosophical."

The many-worlds theory (multiverse) leads to a complete new interpretation of our reality. There is a strong argument for many-worlds theory by David Deutsch. In his book "The Fabric of Reality" [83] he poses the question: "*To those who still cling to a single-universe world-view, I issue this challenge: explain how Shor's algorithm works. I do not merely mean predict that it will work, which is merely a matter of solving a few uncontroversial equations. I mean provide an explanation. When Shor's algorithm has factorized a number, using 10^{500} or so times the computational resources then can be seen to be present, where was the number factorized? There are only about 10^{80} atoms in the entire visible universe, an utterly minuscule number compared with 10^{500}. So if the visible universe were the extent of physical reality,* *physical reality would not even remotely contain the resources required to factorize such a large number. Who did factorize it, then? How, and where, was the computation performed?"* For David Deutsch the physicists answering the question will then either talk some obfuscatory nonsense, or will explain it in terms of parallel universes. Which will be newsworthy. Many Worlds will then become part of our culture. In the year 2001, Shor's algorithm was demonstrated by a group at IBM that factored 15 into 3×5 using the NMR implementation of a quantum computer with 7 qubits [283]. Where were the 128 states represented?

5.2 THE PREFERRED BASIS PROBLEM

Any quantum state is in superposition to some basis. In the many-worlds interpretation the universe splits in a certain preferred basis, but which is the basis? In other words, form which perspective do we look at the quantum state? Without a preferred basis in which to write the universal state, one is free to choose any basis one likes with the result that in different bases there will be different possible worlds.

For example, a quantum state could be described by a basis that represents a superposition of a quantum state. The argument of the preferred basis problem goes like this, imagine in the Schrödinger's cat paradox, after measuring we have a dead cat. Now we close the room and change the perspective of the observable, the basis of the new coordinate system would be represented by the superposition of an axis of dead and an alive cat and the other axis would be also dead and an alive cat but with a different sign [107]. This sounds metaphysical since an observer defines the basis, he is a part of the measurement. Instead of a dead and alive cat, we replace it by a particle with two states represented by two spins. The interpretation of an observer becomes negligible. In many worlds interpretation there are continuous splits and measurements. How does the world know the "correct" basis?

A particle is a part of the world, and each world is an emergent object with a well defined structure [289]. For example, each cell in a human body is a part of the emergent structure of the human body. The world as an emergent structure induces the correct basis that is part of it. With the discovery of quantum decoherence in the 1970s the problem was solved. A quantum system that is perfectly isolated maintains its coherence; it is present in a superposition. If it is not perfectly isolated, during interaction with the environment decoherence happens. The environment measures the system according to its own basis and records the results as a part of one world. The decoherence is a constant process that entangles the environment into a world by the decoherence-preferred basis. The decoherence-preferred basis is defined by the world, the observer is a part of the world. The modern version of the wave function collapse is based on decoherence and leads to the multiverse interpretation of quantum mechanics [37].

5.3 DECOHERENCE IN THE UNIVERSE

Why can we never see cats that are both dead and alive? Why do tiny objects such as electrons and photons exist in a superposition of states but larger objects like footballs and planets apparently do not? Bousso and Susskind's answered this question by alternative realities of the many worlds [37]. We can never see large objects (macroscopic) in superposition contrary to microscopic objects like photons because of decoherence, objects interact with their environment, allowing information about possible superpositions to leak away and become inaccessible to the observer. All that is left is the information about a single state, the information about the other states leaks away and is not accessible. Macroscopic objects tend too quickly become entangled due to decoherence. Decoherence arises from the description of the world by an observer. The environment is defined by what the observer can measure. For the universe there is no environment outside itself, it implies that in this description decoherence cannot occur. How can we solve this problem? The lack of any environment implies that decoherence cannot occur in a complete unitary description of the whole universe. An electron may be in a superpositions. The electron's wave function collapses when the electron becomes entangled with some detector at certain time. Imagine that the electron collapses in some laboratory, the environment in this case is defined by the laboratory. The laboratory is on earth that is part of the solar system, and the environment could be defined as being beyond the orbit of Pluto. In that case the collapse of the system wave function cannot take place until a photon from the detector has passed Pluto's orbit. Since the limitations of the speed of light, this would take around five hours during which the system wave function is coherent [37].

The speed of light defines causal restrictions, called the causal diamond (causal patches), it is the largest region that can be causally probed. No observer inside the universe can ever see more than what is in their causal diamond [37]. The causal diamonds make up the multiverse, the decoherence and the quantum evolution appears inside the causal diamonds. The causal diamonds represent different histories. They split from other universes leading to a global multiverse represented by the many worlds in a single geometry. The multiverse is a patchwork of infinitely many branching histories, the many worlds of causal diamonds.

The system must become irreversibly entangled with the environment inside the causal diamond to give objective meaning to an event. Without revertible decoherence, one cannot claim that anything really happened. Decoherence must be irreversible, and it must occur infinitely many times for a given experiment in a single causal diamond. Things happen, there exist some entanglements that will not be reversed. This is possible in the eternally inflating multiverse, there are infinitely many causal diamonds, however there are only finitely many histories that end in an eternal endpoint [37]. The world is viewed as an unbounded collection of an eternally inflating multiverse. One can view the eternally inflating multiverse global like looking at a tree and seeing all its branches and twigs simultaneously. An observer in the universe sees is as an ant climbing from the base of the tree to a particular twig along a specific route.

5.4 PROBABILITIES IN THE EVERETT MANY-WORLDS THEORY

Every time a quantum experiment with different possible outcomes is performed, all outcomes are obtained. If a quantum experiment is preformed with two outcomes with quantum mechanical probability 1/100 for outcome A and 99/100 for outcome B, then both the world with outcome A and the world with outcome B will exist. A person should not expect any difference between the experience in a world A and B. The open question is the following one: Due to deterministic nature of the branching, why should a rational person care about the corresponding probabilities? Why not simply assume that they are equally probable due to deterministic nature of the branching [113]. How can we solve this problem without introducing additional structure into the many-worlds theory?

5.4.1 Decision-Theoretic Argument

David Deutsch suggested that decision theory can solve this problem [84, 287, 288]. A person identifies the consequences of decision theory with things that happen to its individual future copies on particular branches. A person who does not care in receiving 1 on the first branch A and 99 on the second branch B labels them with a probability 1/2. A rational person that cares assigns the probability 1/100 for outcome A and 99/100 for outcome B.

Decision theory according to Savage is a theory designed for the analysis of rational decision-making under conditions of uncertainty [84]. A rational person faces a choice of acts as a function from the set of possible states to the set of consequences.

Act is a function from states to consequences, the preferences must be transitive. If a rational person prefers act A to act B, and prefers act B to act C, then the same person must prefer act A to act C. This can be summarized by assigning a real number to each possible outcome in such a way that the preferences are transitive. The corresponding number is called the utility.

In the context of the many-worlds the rational person is able to describe each of its acts as a function from the set of possible future branches that will result from a given quantum measurement to the set of consequences [84]. Consequences are the things that happen to individual future copies of the person on particular branch.

It can be then shown that there exists a probability measure p on states s and a utility function U on the set of consequences of an act A so that the expected utility is defined as

$$EU(A) := \sum_s p(s) \cdot U(A(s)). \tag{5.1}$$

It follows that a rational person prefers act A to act B if the expected utility of A is greater than that of B. The behavior corresponds to the maximization of the expected utility with respect to some probability measure [237].

For any two acts A, B, the rational person prefers A to B if

$$EU(A) > EU(B).$$

The person acts if she regarded her multiple future branches as multiple possible futures. Out of this relation one can derive of probabilities for real and irrational amplitudes as proved by David Deutsch [84].

It should be noted that David Deutsch introduced a rational person into the explanation of the corresponding problem. However since the branching is deterministic and no uncertainty is present, how can this rational person justify the application of decision-making? Trying to answer this question may lead to some philosophical difficulties [113, 114, 172]. Instead of rationality we introduce a biological principle that overcome this question.

5.4.2 Homeostasis and the Allostasis Argument

A fundamental property of rational persons is that they prefer certainty to uncertainty. Humans prefer to choose an action that brings them a certain, but lower utility instead of an action that is uncertain, but can bring a higher utility [93].

We use Shannon's entropy as the expected utility in Deutsch's approach [306]. Probabilities in Shannon's entropy function can be seen as frequencies; they can be measured only by performing an experiment many times and indicate us the past experience. Surprise is inversely related to probability. The larger the probability that we receive a certain message, the less surprised we are. For example, the message "Dog bites man" is quite common, has a high probability and usually we are not surprised. However the message "Man bites dog" is unusual and has a low probability. The more we are surprised about the occurrence of an event the more difficult an explanation of such an event is. The surprise is defined in relation to previous events, in our example men and dogs. We say that events that seldom happen, for example, the letter x in a message, have a higher surprise. Some letters are more frequent than others; an e is more frequent than an x. The larger the probability that we receive a character, the less surprised we are. Surprise is inversely related to probability. The logarithm of surprise is the self-information or surprisal. The Shannon's entropy H represents the weighted sum of surprisals and can be interpreted as an expected utility [306]. We can recover the probabilities from the definition of Shannon's entropy. This strengthens David Deutsch's idea even more.

In the next step we introduce the experience of identity derived from homeostasis as a fundamental biological principle. The experience is preformed subconsciously by our brain as a coherent explanation of events in a temporal window [220]. Events with higher surprise are more difficult to explain and require more energy. Before an event happens an explanation has to be initiated, so after the event happened it can be integrated in the present explanation in the temporal window. A rational person may not care about the attached weights during deterministic branching, but our brain machinery cares. This information is essential for the ability to give a continuous explanation of our "self" identity.

Identity is a concept that defines the properties of a rational person over time [221]. It is a unifying concept based on the biological principles of homeostasis [32, 115]. Organisms have to be kept stable to guarantee the maintenance of life, like for example, the regulation of body temperature. This principle was extended by Allostasis [262] for regulation of bodily functions over time. To preform this task efficient mechanisms for the prediction of future states are needed to anticipate future environmental constellations [21, 285]. This is done, because the homeostatic state may be violated by unexpected changes in the future. It means as well that every organism implies a kind of self identity over time [322]. This identity requires a time interval of finite duration within which sensory information is integrated. Different sensor information arrives at different time stamps. The fusion process has to be done over some time window. Similar problems are present during a sensor fusion task in a mobile robot. For example, in visual and auditory perception in humans the transduction of the acoustic information is much shorter than the visual [219]. In it is suggested that in humans a temporal window with the duration of 3 s is created [220]. This window represents the psychological concept of "now" [322]. The consciousness concept of "now" represented by the temporal window is shifted backward in time of the consciousness itself, since a subconsciousness mechanism is required to preform the integration task.

Events with higher surprise are more difficult to explain than events with low surprise values. An explanation has to be possible. When the surprise is too high, an explanation may be impossible and the identity of self could break. The idea is related to the general constructor theory of David Deutsch [86]. The metabolic cost of neural information processing of explaining higher surprise events require higher energy levels than lower surprise events. Fechner's law states that there is a logarithmic relation between the stimulus and its intensity [103]. We assume as well that there is a logarithmic relation between the cost of initiation an explanation of an event and its surprise value. Neuronal computation is energetically expensive [168]. Consequently, the brain's limited energy supply imposes constraints on its information processing capability. The costs should be fair divided into the explanation of all predicted possible branches since the organism will be deterministic present in all of them. A possible solution is given by the Shannon's entropy. The resulting costs of initiating some explanations for action A predicted branches before a split are represented by Shannon's entropy $H(A)$. For the human (subconsciousness) brain it makes sense to choose A to B if

$$H(A) < H(B).$$

since it requires less energy. The probabilities are induced by Shannon's entropy that measures the uncertainty of events [306].

Instead of rationality we introduced the biological principle of Allostasis [262] for regulation of bodily functions over time. To preform this task efficient mechanisms for the prediction of future states are needed to anticipate future environmental constellations [21, 285].

If a quantum experiment is preformed with two outcomes with quantum mechanical probability 1/100 for outcome A and 99/100 for outcome B, then both the world with outcome A and the world with outcome B will exist. The world B is in a less probable, due to its higher surprise it is more difficult to explain than the world A. Every organism implies a kind of self identity over time [322]. An explanation of events has to be possible. When the surprise is too high, an explanation may be impossible and the identity of self could break [304, 306]. For impossible explanations, causality does not exist, and the identity of the self breaks. Only in meaningful causal worlds may personal identities exist. That is why an organism cares about the corresponding probabilities that are related to the self identity over time.

5.4.3 Randomness

The many worlds interpretation leads to a deterministic interpretation of physics without any randomness present in quantum theory. Despite determinism nothing can be predicted by the observer in one universe. The observer follows one path through the diverging branches. He interprets the resolution into one particular outcome as a random collapse, for the observer true randomness exists since he never perceives the splitting.

5.5   BRANCHING REJOINING AND INTERFERENCE

The decoherence causes the universe to develop an emergent branching structure. One can interpret the branching structure either as a tree structure or as parallel different histories [85]. We can assume that many universes are identical in every aspect and then at a certain point they become little bit different by small physical changes. Small effect can break the identity, identical objects become different, and the number of histories increases. The world, which we inhabit is one of a vast number of many worlds, most of which are in practice isolated form one another. A splitting of a *history* in $history_1$ and $history_2$ can be described as

$$history \rightarrow history_1 + history_2$$

Histories can branch and can rejoin, rejoin is time reverse of splitting and indicated as

$$history_1 + history_2 \rightarrow history \wedge interference$$

$history_1$ and $history_2$ merge to one single *history* (single point) and during the rejoin interference is always present. Rejoin is only possible as long as the histories are not entangled with the world in which they appear. A history can spilt and rejoin into one history generating interference

$$history \rightarrow history_1 + history_2 \rightarrow history \wedge interference$$

as long as $history_1$ and $history_2$ are not entangled with the rest of the world, they stay coherent and do not interact with the world [85]. Otherwise, they split into two different worlds and it is not possible to rejoin as $history_1$ and $history_2$ and they become entangled with the world resulting in a split

$$history \rightarrow (history_1 + history_2) \cdot world \rightarrow history_1 \cdot world + history_2 \cdot world.$$

Only in coherent evolution interference appears. During the split no interference is present.

Entangling with more and more objects cannot be undo, it means that the decoherence leads to autonomous independent histories. Observation destroys interference resulting in a split. However there is no psychokinetic effects during the observation.

5.5.1 Electron and Double Slit

Every particle has counterparts in another universes and is interfered with only these counterparts. Interference only in special situations where the paths of the particle separate and converge. For example, the electrons are heading to the same position [85].

A gun that fires electrons and a screen with two narrow slits x and y and a photographic plate. An emitted electron can pass through slit x or slit y. The electron is spited into a superposition of two states that pass at the same time through both slits. Since the electron is unobserved and coherent it is rejoined into a single object at the photographic plate an interference pattern appears. Consequently, the summation rule of classical probability theory is violated due to resulting interference during the rejoin.

On the other hand, if the electron that is in superposition after being emitted from the gun is observed, a split appears into two different histories and without interference.

It follows that the wave-particle duality, all matter exhibits both wave and particle properties can be interpreted as follows. An unobserved electron in a superposition can rejoin producing the interference, an observed electron in superposition splits into two different histories. Since decoherence is present on the macro scale, we notice the interference during the rejoin mostly on the on the micro scale. This is the reason why interference effects are usually weak.

5.5.2 Quantum Cognition

Humans when making decisions violate the law of total probability. The violation can be explained as a quantum interference resulting from the quantum probabilities (amplitudes) [51]. Subconsciously humans assume that the two different histories that lead to the same result cannot entangle with their world since they are present in the future, resulting in the interference that violates the law of total probability [307,308].

During the experiment the participants were shown pictures of faces. They should categorize the person represented on the pictures as good or bad and in the next step decide to act friendly or aggressive. There were two experimental conditions:

- Make a decision after categorizing a face.

- Make a decision without reporting any categorization.

In the second condition of the experiment a decision without reporting any categorization was determined that does not correspond to the correct value as indicated by the law of total probability [308].

There are many more related experiments, like the prisoner's dilemma game.

In the game there are two prisoners, prisoner x and prisoner y. They have no means of communicating with each other. The participants of the experiment were asked three different questions.

- What is the probability that the prisoner y defects given x defects?

- What is the probability that the prisoner y defects given x cooperate?

- What is the probability that the prisoner y defects given there is no information present about knowing if prisoner x cooperates or defects?

In the third question of the experiment a probability value was determined that does not correspond to the correct value as indicated by the law of total probability [308].

Why do humans violate the law of total probability during decision making? We can assume that subconsciously humans model two different histories that a rejoined in the final decision leading to an interference term that violates the law of total probability during decision making. It leads to the assumption that two different histories can rejoin into one history on the macro scale of the world. Also, that the subconscious decision-making process in the brain adapted to this assumption during the evolution when making predictions during the decision process in the future [308].

5.5.3 Quantum Computing

Either the interference appears because the histories are rejoined or the interference cancels the other histories through destructive interference.

The destructive interference plays an important role in quantum computing. Individual objects are represented by qubits, histories represent the states of quantum qubits during quantum computation. In quantum computing we try to control the probability that a system of qubits collapses into a measurement state. We combine alternative histories into one single history [85].

Quantum interference is what allows us to bias the measurement of a qubit toward a desired state or set of states. Quantum are mapped into a superpositions, during the superposition computation in parallel is performed. The qubits can be then mapped in superpositions that have amplitude signs leading to destructive interference of non-result, and constructive interference of the desired state that represents the result of the computation. This desired state is then measured with theoretical certainty. A quantum computer executes parallel computation with the help of multiverse. Few hundred qubits could perform far more computations in parallel then there are atoms in the universe.

Qubits are spitted into superposition, computation in parallel is executed. With the help of destructive and constructive interference the histories are rejoined into one state.

If decoherence during quantum computation appears it is not possible to use the destructive interference and a split appears and measurement quantum randomness will be present.

5.6 STRUCTURE OF THE MULTIVERSE

Common-sense of time divides the time in past moments, present moment, and the future moments [83].

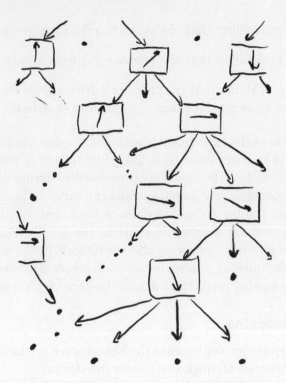

Figure 5.3 Branching structure of the multiverse represented as a tree composed of snapshots. We represent each snapshot abstract as an arrow in a picture with a certain orientation. The branching structure trough decoherence defines the direction of time; it is in one direction and cannot be in the other [83].

5.6.1 Time

Time is not continuous since there are no measurable continuous in quantum physics. Notion of continuous range of possible values of time is just an idealization. The same goes with the direction of time that is a prior assumption based on the branching structure of the multiverse [83]. The branching structure trough decoherence defines the direction of time, it is in one direction it cannot be in the other one, see Figure 5.3. Let us according to David Deutsch analyze the discrete structure of time by representing a moving object. We represent the moving object as a sequence of snapshots [83], see Figure 5.4. It snapshots corresponds to a time frame like an image in a film. The present moment is not moving through the snapshots, it is represented by each snapshot. It is sweeping through pictures like an image in the moving pictures of a film in a cinema. In the snapshot representation there is no distinguishing between the moment, see Figure 5.5.

If we imagine an observer in three-dimensional space, it is him who defines its own coordinate system. Either he moves through space, or he is fixed and the space moves. For another independent observer he moves through space. In the same sense time does not move, it is the observer's consciousness that moves [83]. The consciousness pass from one moment to another one. Our consciousness exists at all our waking

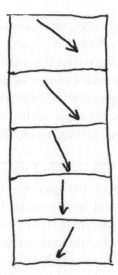

Figure 5.4 We represent the moving object as a sequence of snapshots, which corresponds to a time frame like an image in a film. The present moment is not moving through the snapshots, it is represented by each snapshot; It is sweeping through pictures like an image in the moving pictures of a film in a cinema.

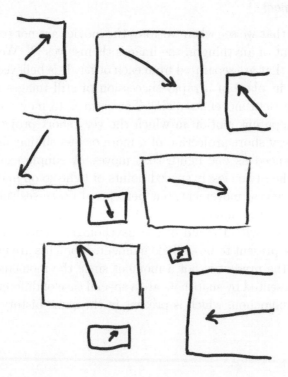

Figure 5.5 Different snapshots. In the snapshot representation, there is no distinguishing between the moment [83].

moments. Objectively there is no present, the consciousness and the mind experiences differences between present perceptions and present memories of past perceptions. It interprets those differences as evidence that the universe changes in time. Humans are accustomed to time being a framework exterior to any physical entity, like an eternal clock. The idea of time flow presupposes the existence of a second time measured by a kind of absolute clock. Time would move with respect to a clock; however, such a clock does not exist, it is an illusion highlighted by the relative theory where absolute time does not exist.

The illusion is reinforced by the usage of time in our society for synchronous reasons. The synchronous time was introduced by the railway timetables, first introduced in United Kingdom 1839 and by a standard railway time in Great Western Railway timetables in 1840, when all their trains were scheduled to Greenwich Mean Time (GMT). Until railway time was introduced no synchronous time existed, local times in English cities could differ by as much as 16 to 20 minutes; in India and North America these differences could be 60 minutes or more.

Flow of time does not exist, only a sequence of moments exists, since no absolute time is present. A snapshot itself is static, but the mind cannot understand it as being static, it is an illusion related to the moving images in a cinema.

5.6.1.1 Beta Movement

The moving images that we see when we go to the movies are not really caused by the continuous movement of anything in the images themselves [8]. We are watching is a series of still images that are separated from each other. It is believed that what makes us perceive motion in place of a rapid succession of still images is a psychological effect, known as the Beta movement [261]. The term Beta movement is used for the optical illusion of apparent motion in which the very short projection of one figure and a subsequent very short projection of a more or less similar figure in a different position are experienced as one figure that moves. A simple example of the Beta movement occurs when two closely spaced points of light go on and off in succession. Though there is no actual movement, our perceptual processes subjectively link the two points into one that is moving.

The common sense with the aid of the psychological effect related to the Beta movement wants the present to be a single moment. Moments are unchanging like the images in the film, the present is not a moment since the moments are unchanging. Different times represented by snapshots are a special case of different universes. Time is an entangled phenomenon, which is present in the same history and perceived by our mind.

5.6.1.2 Past

We think the past is fixed, and the future is open. However, we exist in a moment and the only information about our past is presented to us by our memory that becomes more and more blurry the further in the past we go. We could argue to some extent that the past is open to us as well. We exist in multiple versions, in universes called independent moments. Each version is aware of the other one since they are connected to each other by our memory.

5.6.2 Cause and Effect

We think as well of causes as preceding their effect. All snapshots in a history are glued in the right order because this represents relationships between them that are determine by the laws of physics. Like the Beta movement in which the very short projection of one figure and a subsequent very short projection of a similar figure in a different position are experienced as one figure that moves. The figure must be similar otherwise the illusion of the movement, the time flow is missing [83].

Snapshots have an intrinsic order that is defined by their contents and by the law of physics [83]. The predictability of one event of another does not imply that those events are cause and effects. In a puzzle a piece is missing, we know where to put the missing piece, but it does not mean that the other pieces cause the missing piece. The order of assembling a puzzle is not determined, since the pieces of a puzzle were not disassembled in an order [83].

The law of physics determines what happens from one moment to another, the different moments are glued by the law of physics [83]. Physical laws like the branching structure trough decoherence that defines the direction of moments that are glued, they are glued in one direction it cannot be in the other one.

Events cannot be predicted by each other; no event caused another one in space time [83].

For X to be cause of Y, two conditions must be present, X and Y must happens, if Y did not happen X had been otherwise [83]. Certain events are rare or very common, some events follow other in most cases but is does not mean that they cause them. Low probability events like winning the lottery cannot explain what happened without invoking the existence of many losers.

The causal effect in our perception is an illusion, like the continuous movement in the movies. Causal effect is only defined through different universes, if X causes Y in our universe, in another similar universe X does not happen and Y does not happen either, only then we can state that X causes Y. A snapshot is not determined by previous snapshots of the same spacetime, but by super snapshots of all the other universes. One can make only prediction if one knows what happens in all universes, however we can only perceive one universe [83]. The snapshot description is an oversimplification, since there are snapshots inside the snapshot, multiverses that exists and disappear inside a universe like for example, a quantum computation with a measurable result. In reality we cannot define clear boundaries between single snapshots or super snapshots, see Figure 5.6.

We can make only probabilistic predictions in our reality of our own experience, even when the multiverse is strictly deterministic [83].

The multiverse does not exist in external space nor time since there is nothing more beside it. That is also why there are no timestamps on the snapshots. Other times in our universe are distinguished from other universes only from other universes from our perspective since they are closely related to our laws of physics. We cannot

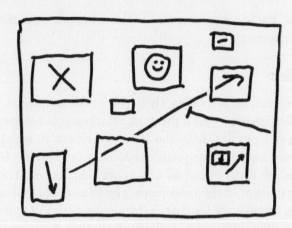

Figure 5.6 The snapshot description is an oversimplification since there are snapshots inside the snapshot, multiverses that exist and disappear inside a universe, e.g., a quantum computation with a measurable result. In reality we cannot define clear boundaries between single snapshots or super snapshots.

glue the snapshots arbitrary, but only in a way that reflects the relationship between them that are determined by physical laws. Some regions of the multiverse, and some places in space, fall in chains, each member determines the other one as a good approximation and is interpreted by us causality. However, this approximation breaks down if one examines the snapshots in more detail like looking into the measuring paradox [83, 85]. According to quantum physics the snapshots are not in any order. The sequential property of time does not exist. Sequential order only begins to exist shortly after the Big Bang. During the Big Bang the backdown of sequence of time was present, similar breakdown is present in in black holes and the final collapse of the universe (Big Crunch) [83].

All fiction that does not violate the laws of physics is a fact since it represents a valid history in the multiverse. [83, 85] A puzzling remark about dreams. In relation to the multiverse is the following one, dreams are not real for us, however we can assume that everything that is possible exists so possible narratives of dreams exists in other universes. Our universe is an emergent feature of the multiverse, the same is valid for mind.

5.6.3 Multiverse and the Turing Machine

Digital physics states that the universe equivalent to a Turing machine. However digital physics are incompatible with quantum physics. A Turing machine in a universe cannot simulate the spacetime events.

However, we could simulate such events by a quantum computer as suggested by Richard Feynman.

The law of physics determines what happens, but in a universe, we can make only probabilistic predictions. A snapshot is not determined by previous snapshots of the same spacetime, but by super snapshots of all the other universes. The multiverse is deterministic, as a result it can be described by a Turing machine. The multiverse cannot be simulated since there is no external space nor time in which a Turing machine could be present.

A Turing machine can only describe the whole multiverse, it cannot describe independent parts of it. It cannot describe a universe and the spacetime in which we are living.

5.7 MULTIVERSE AND THE MIND

5.7.1 Quantum Suicide

Quantum suicide is a thought experiment [289]. A gun is rigged to a machine that measures the spin of a quantum particle. Each time the trigger is pulled, the spin of the quantum particle is measured with 50 percent chance. If the particle as spinning in a clockwise motion the gun will fire, otherwise it will not fire. A man in the closed room pulls the trigger and the gun does not fire. The man will continue to pull the trigger again and the gun won't fire. He'll continue this process for eternity, the gun never fires. After every iteration only one of the two experimenter superpositions that is alive one is capable of having any sort of conscious experience. The experimenter survives all iterations of the experiment. After any number of iterations, the survival is a physically necessary; hence, the notion of quantum immortality.

Eugene Shikhovtsev's biography of Everett [252] states that "Everett firmly believed that his many-worlds theory guaranteed him immortality: his consciousness, he argued, is bound at each branching to follow whatever path does not lead to death."

Tegmark suggested that what makes the quantum suicide work is that it forces an abrupt transition [275]. The flaw in that reasoning is that dying is not a binary event as in the thought experiment; it is a progressive process, with a decreasing consciousness. In most real causes of death, one experiences such a loss of self-awareness.

5.7.2 Many-Minds

The observer's mind should be not in a superposition with himself. Because of this, in the Many-minds interpretations it is argued that the distinction between worlds should be made at the level of the individual observer [5]. The human bodies are in a superposition, but the minds have definite states that are never in superposition. When someone answers the question about which state of a system an observer is, he must answer it with complete certainty. When an observer becomes entangled during a measurement of quantum system, he will be a part of a larger quantum system that splits. Each possible branch has its own mental state of the brain. After the split the mental states of the other branches become inaccessible [5, 82]. Each observer has a continuous infinity of minds. The observer's body evolves in the usual deterministic way, but each mind jumps randomly to a mental state corresponding to one of the Everett branches that is produced in each measurement-like interaction. Each mind follows a branch, a path in the multiverse and becomes different in a world with a particular outcome. During the superposition the observers' eyes, body and brain are seeing both branches at the same time. The mind however, stochastically chooses one of the directions, and that is what the mind sees. The probability of realizing a specific quantum measurement is directly correlates to the number of minds they have where they see that measurement.

The world does not split, it has a tree like structure of the multiverse [5,82]. However, we cannot see this tree structure, since our mind splits and follows each brunch individually. Lockwood introduces the term Mind to denote the multiple entity that is having all the branch tree structure experiences, and the term mind for an entity that is following a branch. In each layer, the experiences correspond with each other and with the physical objects that they are experiences. There is no experience of any other layer, except indirectly, through interference phenomena. The Mind sees all branches, but quantum entanglement connects them in layers. Each of these layers is a universe, the layers of the multiverse are parallel universes. This approach requires a strong mind-body dualism. I could drink red wine. Quantum theory implies that vast numbers of other experiences of mine, including the experience of drinking white wine or water. The reason why I do not have an experience of having all those experiences is as Lockwood puts it; "..*the gaze of consciousness to a kind of 'tunnel vision' directed downwards in the experiential manifold. We cannot look 'sideways' through the manifold, any more than we can look 'upwards', into the future.*" We assume that our experiences have a single-valued history, but this is not the case.

The multiverse is viewed from the perspective of the mind [82]. Other theories of the multiverse assume that the layering structure has an observer-independent basis. In the many minds theory, the tree structure of the universe is present, only we are unable to perceive it. The Mind looks at a tree and seeing all its branches and twigs simultaneously. The mind sees is as an ant climbing from the base of the tree to a particular twig along a specific route, it follows a layer.

This solution comes at the price of introducing additional structure into the theory, including a genuinely random process in a deterministic universe.

5.7.3 The Explanation Hypothesis and the Multiverse

5.7.3.1 Identity

What makes me the same person over time? Is it the persistence of my body? If parts my body are replaced, like liver, lungs, or heart, I'm still the same person? If it is not the body that is essential to being the same person, then it must be something related to my memories and my character. But these come and go over a lifetime; most people remember very little of their early childhoods, and some people have trouble remembering what they did. We exist in a moment and the only information about our past is presented to us by our memory that becomes more and more blurry the further in the past we go. The past is open to us the further we go. What defines our identity, what stays the same in future and past?

Identity is a concept that defines the properties of a rational person over time. It is a unifying concept based on the biological principles of maintaining stability and the regulation of bodily functions over time called allostasis [262]. Efficient mechanisms for the prediction of future states are necessary to anticipate future environmental constellations [285], [21]. This is done by causally consistent explanation of events that maintain self-identity over time, leading to the psychological concept of "now."

5.7.3.2 The Moment "Now"

How can we define the concept of "now" according to the many-worlds interpretation? The common sense with the aid of the psychological effect related to the Beta movement defines the psychological concept of "now." The present moment is not moving through the snapshots, it is represented by each snapshot. The branching structure trough decoherence defines the direction of time, it is in one direction it cannot be in the other one. If an explanation and the psychological concept of "now" is not possible due to a lack of causality, the identity of the self may break. The implication is that that only in meaningful causal worlds personal identities may exist. All snapshots in a history are glued in the right order because this represents relationships between them that are determine by the laws of physics.

5.7.3.3 The Task

The consciousness is present in the "now', it recovers the information what was before with the aid of information that is stored in the memory and tries to predict what will happen later. The main task of the brain is to provide a causally consistent explanation of events and to decide, which actions keep the homeostatic state in the future. Probabilities are interpreted as surprises, the lower the probability, the higher the surprise, the less common the branching of the world is, more energy should be allocated in the process of temporal integration of less common worlds. The mind facilitates the maintenance of self-identity over time that leads to an evolutionary advantage through energy efficient preparation of temporal windows that allow an integration of a temporal explanation of the moment.

5.7.3.4 Subconsciousness and Superposition

The consciousness concept of "now" represented by the temporal window is shifted backward in time of the consciousness itself, since a subconsciousness mechanism is required to perform the integration task. Different times represented by snapshots are a special case of different universes. The subconsciousness perceives the superposition of states, it perceives our eyes, body at both branches at the same time.

5.7.3.5 Things Happen

The temporal window is shifted backward in time it represents time as an entangled phenomenon of one history. It is allocated after the branching structure trough decoherence that defines the direction of time and corresponds to the entanglement process by the subconsciousness process of the brain, see Figure 5.7.

Only after the decoherence anything really happened. Different causally correlated snapshots are fused to temporal window with the duration of 3 s [220]. The body and the mind represented by the temporal window splits to one of the

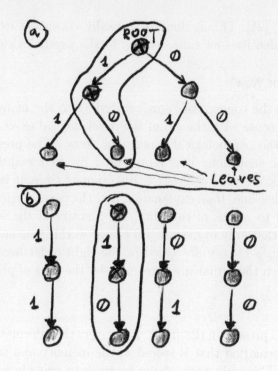

Figure 5.7 The temporal window is shifted backward in time representing time as an entangled phenomenon of one history. It is allocated after the branching structure trough decoherence that defines the direction of time and corresponds to the entanglement process by the subconscious process of the brain. The tree expands and the leaves represent the moments of each history before decoherence. The crosses represent the decoherence. Lower part represent the resulting different histories.

Everett branches that is produced in each measurement-like interaction. There is no randomness involved like in the many-minds approach.

5.7.3.6 *Independents Moments*

Each branch corresponds to a temporal window. The temporal windows are similar, they are independent and not connected to each other. The temporal windows of the same history are as well independent, however they are connected through memories. We in different times are different persons only related to each other by our histories. During the branching process we are similar, however with continuous additional branching the similarity becomes less, different personalities will be present. This also happens with us during our history in time, with continuous time we can become different personalities. However, our brain tries to maintain an identity through the explanation process with the aid of memories. We exist in multiple versions, in universes called independent moments. Each version is aware of the other one since they are connected to each other by our memory. We are defined by our branching history.

5.7.3.7 Sensing the World

The temporal identity is more related to human senses than to an algorithmic device. From this perspective, artificial intelligence models human cognition but not the mind or consciousness. The consciousness can perceive the world through a temporal window with the help of our body and the intelligence. We humans live on land, to explore life under the water, we need diving equipment. Our body's capabilities are expanded so that we can perceive the world under the water. And it is the same with consciousness. To perceive the world, one needs equipment, in the case of humans the equipment is the human brain together with the human body of which brain is a part of. The bodies together with the brain allow the consciousness to perceive the world. Different bodies offer the consciousness different images of the world in different qualities. What does consciousness perceive with the insect's body?

5.7.4 Dualism and Free Will

The self-identity and self-consciousness inside the temporal window correspond to a combination of qualia that indicates the quality of a causal story. Qualia are nonphysical and are perceived by our mind into the experience of self-consciousness. The consciousness represented by connected qualia and the brain are two different things. The consciousness is the unchangeable nonphysical part of us, it interacts with the brain only in one direction: The consciousness senses the brain trough the temporal window but does not make decision known to the brain, it is the **unchangeable** part of us. It is nonphysical and it senses the world. Every mental state is identical with some state in the brain. Despite the fact, we indicate how free will can be present that defines our personality.

Incompatibilism states that a deterministic universe is logically incompatible with the notion that people have free will. Then again, it is supposed that non-deterministic quantum mechanics plays an essential role in the understanding of the human mind and free will. It is assumed that the mind is non-algorithmic and thus incapable of being modeled by a computer. We assume, first, that the universe is deterministic and, second, that the mind interacts with the brain only in one direction: The mind senses the brain but does not make decision known to the brain. Despite these two major constraints, we indicate how free will can be present

5.7.4.1 Multiverse Metaphor: Library of Babel Metaphor

Someone is reading a book about a certain hero and his adventures. During the process of reading, he identifies with the hero and lives through his ups and downs. Reading is a complex cognitive process, and it cannot be described by a simple function. It requires high cognitive functions for textual interpretation and decoding into a cognitive representation and comprehension. The brain preforms these functions; it decodes and interprets the story, but it does not create it. However, an illusion that the reader is the hero described in the book and that the reader is making the decisions may exist. Let us perform a thought experiment. A human being's life, with all of his decisions, is recorded and transcribed into an enormous book called the book of life. A demon allows the human being to live again; however, he would erase all of his memories. The new life of the human consists of reading the previously recorded book of life. Does this mean that, during the reading, the human being is making his free choices or not? One can argue that the person makes free choices because the book describes free choices in a free world and these choices are those of the person. One can argue that, to write such a book, one must live in a free world. The book is a description of a real free life. By reading it, one cannot distinguish between the moment in which one is making choices and reading the book of life [304].

What if some demon knew about the person's free choices and wrote the book of life without the person actually having lived and having made the choices at all? Is it possible that a demon can predict free choices?

In Borges' story, the universe consists of an enormous library of an indefinite and perhaps infinite number of books, see Figure 5.8. The books contain every possible ordering of just a few basic characters. Most of the books are completely useless

Figure 5.8 Jorge Francisco Isidoro Luis Borges Acevedo (1899–1986) was an Argentine short-story writer, essayist and poet. "The Library of Babel" is a short story by Borges conceiving of a universe in the form of a vast library containing all possible 410-page books of a certain format and character set. The books are random and most of them are meaningless. The books contain every possible ordering of just 25 basic characters (22 letters, the period, the comma, and space). The library must also contain every coherent book ever written, or that might ever be written, and every possible permutation or slightly erroneous version of every one of those books. The story was originally published in Spanish in Borges' 1941 collection of stories.

Figure 5.9 Example of 144 random binary images of size 10×10.
$2^{100} = 1267650600022822940149670320537$6 different images exist. This number is much higher than the estimated number of atoms in our universe, which is $10^{28} = 1000000000000000000000000000000$0. However, we could represent this number by a superposition of 100 qubits in a quantum computer.

to the reader and have no meaning. An example of the variation of the 23 letters: *yxnburygapbzcmceolnyvhlai, gj. rlzpkewaiwpmgvgunbuptv. rgajmlrskjqy. udsbkfyogedjflx kvuebaz,kiunppuounvlzqodicnlntfjvbusac,xk z mskhegwocmp im,ecvwgbqndwx **multiverse** u. dp.bv, krhtsbyhsxsqw mghiqnllj.fzrdiwmvisydip, qmi. lbaay. zdc, r.h. jaelgjy. p, cilqjdu. yuqszejohcwm. hjyypuaubanjqpvumjtkzeiuyvwpnbxlhvztcv, rriszwobmrdmlkdcnee wvc zjz,n. p. qpbpflory, pwmlm. gwkyauzxl ig. yysn. imkiaoijzvmxolwargljc mmepvmzqcsj.*, see also Figure 5.9. In a language some letters are more frequent than others and some combination of letters are less probable.

Among all of these meaningless books, there are all the books that are or ever will be written [304]. These meaningful books define causality; in other words, causality is represented by a meaningful book. These include all possible biographies of any person and translations of every book into every language. By chance, a person will find his book. By reading some noise, we become nothing. We are defined by our book of life, as we are defined by our human body. Do we make free choices? Yes, we do because we cannot make any distinction between reading our book of life and living in a free world [304].

A person should not expect any difference between reading book A and book B, but the person reading book A is a different person from the person reading book B. We are one person described by the book that we are reading or, in the language of quantum physics, by the path in the multiverse that is correlated with our free choices [304, 306].

All possible lives, personalities that do not violate the laws of physics are facts since they represent a valid history in the multiverse. We are one of this possible personalities. Our choices are defined by our personality explained to us by our mind, our consciousness is the unchangeable part of us. It is nonphysical and it senses the world. It does not interact with the brain in the other way beside it.

5.7.4.2 4321–A Novel

There are billions similar persons, but they are not us. An example of diverging personalities is described in the novel "4321" by Paul Auster. Archie's lives start the same. Archie falls in love with the same girl, but then it begins to diverge. The book is divided into four parts, which represent the different versions of Archie's life. He grows up with the same middle class parents, as well as many of the same friends, including his girlfriend. However, the relationships change with each Archie version, his lives take very different paths. Depending on the version of his life, Archie experiences various identity issues.

CHAPTER 6

Conclusion

INCOMPATIBILISM states that a deterministic universe is logically incompatible with the notion that people have free will. Then again, nondeterministic quantum mechanics are assumed to play an essential role in the understanding of the human mind and free will. The mind is assumed to be non algorithmic; thus, it is incapable of being modeled by a computer.

6.1 DUALISM AND ETHICS

Dualism represents a philosophical and theological speculation [278]. A philosophical argument on either side is not scientific proof. The dualistic interpretation of mind allows the presence of free will and a definition of humans that goes beyond the mechanistic. However, in addition to the main problem of how the soul interacts with the human body, some ethical problems are suggested. Dualism implies that reality consists of two fundamental types of existence that cannot be reduced to each other, such as spirit and matter. Ethical dualism posits the existence of two mutually hostile entities, one representing the origin of all Good and the other, of all Evil. Ethics are based on the conflict between two antagonistic forces, like light and darkness or Good and Evil [278]. Ethical dualism targets different groups of people, such as the nobility, the bourgeoisie, and many others. All these cases have been and are characterized by injustice. This situation has pushed people who belong to a certain group to associate and link injustice entirely with the group that is perceived to have caused it. The Christian doctrine of Original Sin affects all humans equally. Evil cannot be seen as the exclusive domain of a determined class or group of people. The ethical view of dualism exists only in the mental dimension of human beings. Evil arises in the world through people's wrong choices and actions [278]. Since our soul exists forever, our human life on earth is less valuable. Punishment by death is not so bad, since the soul continues to exist.

Modern dualism is based on quantum physics. Quantum mechanics can be used to argue that an individual can change the world by thinking about it. For example, quantum healing is a pseudoscientific mixture of ideas indicating that quantum phenomena govern health and wellbeing. Quantum healing undermines genuine science and discourages people from engaging with conventional medicine.

DOI: 10.1201/9781003244547-6

6.2 COMPUTATIONALISM AND ETHICS

According to computationalism, human minds are epiphenomenal only [79,191]. Explaining the mind as an emergent property of a complex system is per se not a scientific theory. It is a hypothetical speculation; the hypotheses are not based on empirical evidence. A theory belongs to empirical science if and only if it contradicts possible experiences, i.e., it can, in principle, be falsified by experience, called the falsifiability criterion [222]. Computationalism is increasingly taking over the space in society that religion formerly occupied. Computationalism has arisen with its own truth, which is far removed from the truth of the humanities. A theological truth is unrelated to a physical truth. We stop asking questions that science cannot answer; we ignore them. We ignore borderline situations that arise from antinomies, from situations of the struggle for survival, and from the experience of death [143]. Computationalism is seen as an offspring of the sciences. Knowledge is believed to be derived from the help of the sciences, where it finds its justification, since it starts from an "objective" truth [177,178]. The natural sciences pretend to be objective; therefore, according to many, they are closer than are the humanities to the objectivity that corresponds to truth. However, computationalism is not a scientific but a metaphysical hypothesis.

Descartes assumed that animals are machines. The suffering of an animal is produced entirely by a material mechanism. Since animals lack minds, they are machines and cannot experience pain. Dissecting a live animal is no different than disassembling a mechanical clock. No suffering is experienced and no moral issue obtains [321]. Only humans have minds, and only humans can suffer. Ethics is applied only to living creatures with minds. Computationalism states that the human mind is an information-processing system, and that cognition and consciousness together are a form of computation. Humans are machines, like Descartes's animals. If so, what would be the foundation of morality and ethical actions? Dissecting a live human is no different than disassembling a mechanical clock. Ethical actions cannot be justified by science, as they are not rational. The Ten Commandments are derived from God and not from science. Talking rationally about ethics and ignoring feelings is erroneous. Doing so assumes that ethics could be built from reason, thereby evading the responsibility imposed by one's own conscience. According to Dennett, ethical behavior is based on reason and logical actions, so moral people who cooperate will be more successful than immoral people who do not cooperate, resulting in a beneficial mutual arrangement. However, ethical actions cannot be logical because if one acts logically, one fails [23]. As Zygmunt Bauman states: *A pious old man, traveling with a donkey laden with sacks of food, met a beggar on the street. The beggar asks for some food. "Wait," said the old man, "until I've untied the sacks." But before he could do that, the beggar was overcome with hunger and fell dead on the ground. Then the old man prayed and said: "Punish me, O Lord, for I have neglected to save the life of my fellow man!"* [23].

From 1933 to 1945, Nazi scientists and doctors at preeminent German research institutes researched diligently before determining that Jews and Roma were subhuman. Like all kinds of vermin, they should be eradicated, or they could be used, like Descartes's animals, similar to other beings with lesser minds. Since the Jews and Roma were not full human beings, doing so was not wrong, and eliminating them and their culture was perfectly normal. The liberation of reason from feeling, of rationality from normative constraints, and of efficiency from ethics has been the scientific program since the beginning. In practice, however, the sciences, and with them the technology they produce, was often not based on the scientific method and was transformed into an instrument of unscrupulous rulers. The dark and contemptible role of science in the enactment of the Holocaust has both direct and indirect aspects. Science indirectly paved the way to the Holocaust because it undermined the authority and cohesiveness of normative ideas, especially religion and morality [23]. Science participated directly through the formation of racial theories. After [44], [23] presents an example of an apparently objective, scientifically technical discourse without any morality, for morality arises from a discourse other than that of technology. It has no place in this discourse. The notes of the expert *Willy Just* were analyzed, who suggested improvements for a gas van in which people were gassed. The suggestion related to shortening the loading area of the gas van [23]. *"A shorter, fully loaded vehicle could work much faster. The shortening of the rear structure would not adversely affect the load distribution and would not overload the front axle either, because the load distribution is automatically corrected by the fact that the freight moves towards the rear door during the operation and remains there. Since the supply nozzle rusts quickly when exposed to liquids, the gas should not be introduced from below but from above. To facilitate cleaning, an eight to twelve inch opening in the bottom of the chamber would be beneficial, with a removable cover on the outside. The bottom should slope towards the center and the cover could be fitted with a small strainer. All the liquids would collect in the middle of the chamber. The thin liquids might spill out during operation; thicker liquids could be eliminated with the water jet later."* As a specialist in truck construction, he saw in his inner eye the load movements and not the prisoners' struggle for air; he was thinking of thin and thick liquids, not human excrement and vomit. The fact that the cargo consisted of suffocating people who were no longer in control of their bodies did not alter the technical challenge [23].

These considerations were based on scientific considerations of utility and its enhancement. Since science knows no ethics, the question did not arise of the legality of the action. Therefore, Eichmann's defense argued that his actions had no intrinsic ethical dimension and were therefore not immoral [23].

Computationalism has a clear implication on human society by downgrading humans to complex machines with several resulting ethical implications.

Another consequence of computationalism is the implication that machines that look like and behave like humans have the same rights as humans [192]. The Hitchhiker's Guide to the Galaxy, a comedic science fiction novel by Douglas Adams (1952 - 2001), mostly poked fun at scientific advances, such as the artificial personalities built into the work's robots. Adams predicted some concepts that have since become reality. The fictional Sirius Cybernetics Corporation describes robots as "Your plastic pal who's fun to be with" (see Figure 6.1). Today, such robots exist; they are called companion robots. A companion robot is created for the purposes of creating real or apparent companionship for human beings. These robots have humanlike faces with

Figure 6.1 "Your plastic pal who's fun to be with," is the description of the robot by the marketing department of the Sirius Cybernetics Corporation.

eyes, mouths, and noses. The function of those artificial faces is to communicate with humans through facial expressions that indicate emotions. Even though the robots simulate emotions, it is easy to be fooled and assume that the robots feel real pain and happiness and have a mind of their own (see Figure 6.2). Little children assume that their teddy bears live and feel. They project their own minds on lifeless toys. They are an illusion of the presence of a mind. Both companion robots and teddy bears help humans manage life problems, such as loneliness or fear. However, despite the positive effects they have on human life, such constructs do not have minds. Computationalism supports the idea that robots might someday become conscious, since humans are complex machines as well. Thomas Metzinger proposed a global moratorium strictly banning all research that directly aims at the risks accompanying artificial consciousness machines [192]. Humans are machines like robots. Like machines, they are defined by their functions; they will be dehumanized and lose their human rights.

6.3 MULTIVERSE DUALISM AND ETHICS

Socrates teaches that a man must know how to choose the mean and avoid extremes to the greatest possible extent. In ancient Greek philosophy, especially that of Aristotle,

Figure 6.2 A companion robot is a robot created for the purposes of creating real or apparent companionship for human beings. These robots have human-like faces with eyes, mouths, and noses. The function of those artificial faces is to communicate with humans through facing expressions that indicate emotions. Even the robots simulate emotions so it is easy to be fooled and assume that the robots feel real pain and happiness and have minds of their own.

the golden mean or middle way is the desirable middle between two extremes, one of excess and the other of deficiency.

Aristotle does not support Platonic dualism. He also does not support materialism, which is currently embodied in computationalism. Aristotle seeks a middle course between two extreme viewpoints - materialism and dualism - by noting that these are not the only available options. The explanation hypothesis and the multiverse represent the golden desirable middle way between two extremes of computationalism and dualism. The mind and the brain are two different things, as Descartes assumed, but the mind in the multiverse interacts with the brain only in one direction: The mind senses the brain but does not make decisions known to the brain.

According to multiverse theory, there should be a universe in which every possible mind exists, and each mind corresponds to a personality. Causality is related to a meaningful explanation. For impossible explanations, causality does not exist, and the identity of the self fractures. Only in meaningful causal worlds can personal identities exist. We are one person described by the meaningful path in the multiverse that is correlated with our free choices.

Self-identity and self-consciousness inside the temporal window correspond to a combination of qualia that indicate the quality of a causal story. Qualia are non-physical and are perceived by our mind, forming the experience of self-consciousness. Consciousness is represented by connected qualia. Consciousness is the unchangeable nonphysical part of us. It interacts with the brain only in one direction: Consciousness senses the brain through the temporal window but does not make decisions known to the brain; it is the unchangeable part of us. It is nonphysical and it senses the world. The unchangeable part of us is related to what Neoplatonism calls the eternal immortal world-soul that is part of our mind. The immortal world-soul does not change and is the part of our mind that feels. Something eternal is always there. It is timeless and cannot change. Consciousness is not an emergent property of the material world.

All possible lives, personalities that do not violate the laws of physics, are facts since they represent a valid history in the multiverse. We are one of these possible personalities. Our choices are defined by our personalities, which are explained to us by our minds. Our consciousness is the unchangeable part of us. It is nonphysical, and it senses the world. It does not interact with the brain in any other way [304].

The explanation hypothesis and the multiverse are based on the many-worlds theory (multiverse) or many-worlds interpretation (MWI) [96]. These theories lead to some philosophical questions that, for some scientists, create an uncomfortable feeling, like what is a personality and what is causality.

6.4   OTHER MINDS

How do I know that other beings with minds exist? All others apart from me are possibly automata, zombies, or non feeling individuals who behave similarly to me. The analogical argument cites similarities between two beings and uses this as support for concluding that further similarities may be assumed to exist. Other beings have bodies like me, and they exhibit acts that are possibly caused by feelings. The idea that the mind of another is not directly observable is expressed in the fifth century AD by St. Augustine: "*For even when a living body is moved, there is no way opened to our eyes to see the mind, a thing which cannot be seen by the eyes.*" Thomas Nagel asks how one person can understand the attribution of mental states to others [197]. The philosopher Ludwig Joseph Johann Wittgenstein (1889–1951) writes in Section 302 from Philosophical Investigations [317]: "*If one has to imagine someone else's pain on the model of one's own, this is none too easy a thing to do: for I have to imagine pain which I do not feel on the model of pain which I do feel. That is, what I have to do is not simply to make a transition in imagination from one place of pain to another. As, from the pain in the hand to pain in the arm. For I am not to imagine that I feel pain in some region of his body?*" What are others? Humans tend to attribute minds to humanlike robots with very limited AI and discuss whether such robots should be given moral consideration. Humanlike social robots are special to humans: They are capable of limited decision-making and learning, can exhibit behavior, and interact with people. A controversy involves whether to ascribe legal personhood to robots [258]. Humans tend to develop emotional bonds with robots, attribute human characteristics to them (anthropomorphizing), and ascribe intentions to them [121]. These phenomena can be seen in the social humanoid robot Sophia's being granted Saudi Arabian citizenship in 2017 [211]; see Figure 6.3. This phenomenon occurs only because robots look like us. No one attributes mind to an intelligent refrigerator.

Figure 6.3 Humans tend to develop emotional bonds with robots, attribute human characteristics (anthropomorphizing), and ascribe intentions to social robots [121]. This can be seen in the social humanoid robot, Sophia, being granted Saudi-Arabian citizenship in 2017 [211].

Human tendency to attribute minds to robots, and the argument to give them moral consideration is truly disturbing.

When we interact with animals, we intuitively read thoughts and feelings into their expressions and actions. We suppose that they have minds like ours. How different are the minds of animals to those of robots?

Robots can exhibit high levels of cognition, but cognition and mind are different terms. Cognition is closely related to human intelligence and can be simulated by computers and robots. On the other hand, mind is vaguely defined involving consciousness, a combination of cognition and emotion, and unconscious cognitive processes.

Consciousness remains puzzling and sometimes controversial. It is sometimes synonymous with the mind, and, at other times, an aspect of it. Aristotle claims that, to perceive anything, a person must perceive his or her own existence [11], suggesting that consciousness entails self-consciousness. Self-consciousness can be understood as an awareness of oneself, as a separation between oneself and the world. A not self-consciousness animal can be conscious. It does not make a separation between itself and the world. A little child perceives the world as an entity. Then, over time, a shocking discovery arrives: it and the world are two distinct entities. Life is not

Figure 6.4 When we interact with chimpanzees, we intuitively read thoughts and feelings and suppose that they have minds like ours.

eternal. We assume that a high level of cognition is required for consciousness to become self-consciousness. At what age can children be credited with having self-consciousness? It has been claimed, most forcefully by Gallup and colleagues, that the capacity to recognize oneself in the mirror is a marker of self-consciousness [148]. Is self-consciousness present beyond humans? When we interact with animals, we intuitively read thoughts and feelings and suppose that they have minds like ours (see Figure 6.4). How different are human from animal minds? Does an intelligence derived through an entirely separate provenance from ours have a mind? In "Other Minds," Peter Godfrey-Smith [112] considers octopuses and their minds. Octopuses are different, their brains are distributed; their arms harbor nearly twice as many neurons as the central brains; they have three hearts. Neural loops may give the arms their own form of memory. Some octopuses are shy. Some are confident. They have personalities. They recognize human faces and become fond of certain people, yet at others they squirt disdainful jets of water (see Figure 6.5). Octopuses represent "an independent experiment in the evolution of large brains and complex behavior." Neither language nor a worldview is needed to measure intelligence. Godfrey Smith disagrees with the idea that consciousness suddenly emerged from unthinking matter; it built up in small steps with separate capabilities for perceiving the world [112]. He asks, "*Does damage feel like anything to a squid? Does injury feel bad to a lobster or a bee? Well, insects do not groom or protect injured parts of their bodies. But injured crabs, shrimp and octopuses do. Injected with a chemical thought to spark pain, zebra fish prefer water with a dissolved painkiller; so yes, fish feel pain.*" Primordial emotions such as pain, hunger, or thirst do not require a worldview; they comprise a form of consciousness.

Figure 6.5 Octopuses are different as their brains are distributed, their arms harbor nearly twice as many neurons as their central brain, and they have three hearts. Neural loops may give the arms their own form of memory. Some octopuses are shy, and some confident, as they have personalities. They recognize human faces; they become fond of certain people, yet they squirt disdainful jets of water at others.

Animal behavior is steered by qualia that represent the world: the good and the bad. Rules can be generated only from qualia, since no other objective truth indicates what is good or bad for the organism. Still, we do not know what qualia are.

Animals have minds like humans. Their life stories can be beautiful and poetical. For example, read "The Jungle Book" (1894) by Rudyard Kipling. Robots and computers do not have minds since they do not interact with the multiverse and do not experience qualia.

6.4.1 Other Civilizations

The Earth has minds other than ours: those of the animals. Does our universe have any other civilizations? We cannot communicate efficiently with animals. We are not interested in doing so. How, then, can we expect to communicate with other life forms on other planets? We will only be able to communicate with technical civilizations such as ours. We could confirm such a technical civilization by receiving signals they emit or by seeing "miracles." Seeing "miracles" describes phenomena that are inexplicable from an astronomical point of view [239]. If we received signals, would we know that they are from other civilizations?

In October 2017, a mysterious cigar-shaped object was spotted tumbling through our solar system by the Pan-STARRS 1 telescope in Hawaii. The object, nicknamed 'Oumuamua, was a new class of "interstellar object" neither a comet nor an asteroid. 'Oumuamua, was 10 times as long as it was wide and traveled at speeds of 196,000

Figure 6.6 The object nicknamed "Oumuamua" was a new class of "interstellar objects," neither a comet nor an asteroid. Abraham Loeb, professor and chair of astronomy, and Shmuel Bialy, a postdoctoral scholar, both at the Harvard Smithsonian Center for Astrophysics, raised the possibility that 'Oumuamua, which is 10 times as long as it is wide traveling at speeds of 196,000 mph, might have an "artificial origin," maybe an operational probe sent intentionally to Earth's vicinity by an alien civilization. The object had an excess acceleration and an unexpected boost in speed as it traveled through our solar system. It could be a natural object or a space probe; we do not know and maybe will never know for sure.

mph. Abraham Loeb, Professor and Chair of Astronomy, and Shmuel Bialy, postdoctoral scholar, both at the Harvard Smithsonian Center for Astrophysics, raised the possibility that 'Oumuamua **might** have an "artificial origin." It was perhaps an operational probe sent intentionally to Earth's vicinity by an alien civilization [33]. The object demonstrated excess acceleration and an unexpected boost in speed as it traveled through and out of our solar system (see Figure 6.6).

We should already be receiving signals from the star's vicinity, but it seems that we are not doing so. Why? Is it related to the short life span of civilizations? If the average life span of a civilization is less than twenty thousand years, then the failure of our attempts is understandable. Could it be that civilizations develop only to become extinct after a short time by self-destruction through wars? Our own civilization has not yet overcome this problem. Quite possibly, in most parallel universes, our civilization is destroyed through wars. Or could it be that, of a billion planets in the galaxy, only Earth has produced a civilization? The probability is extremely low of the occurrence of humans. It is based on a chain of phenomena that themselves have extremely low odds of occurrence, such as throwing a hundred dice to roll a hundred (i.e., all the dice roll a one) or dissolving an ice cube in lemonade only for it to appear again. Organisms could only have emerged on a planet that had a large and lonely moon, causing ebbs and flows. The dinosaur extinction was a sudden mass result of a comet's impact on the Earth. The growth of primitive man's brain was accelerated by the negative environment of the glacial period. We do not know what is accidental and what are the inevitable consequences of the development of mind and intelligence [170]. Whales resemble fish, but they previously lived on the land. Whales are intelligent, but they have not built any civilizations since they have no need to do so. Dolphins have an amazing intelligence in comparison with other oceanic creatures, but they have not developed any technical civilization. Planets that "favor" life through climactic and geological stabilization can be homeostatic areas upon which no civilizations develop, since there is no need for life to continually adapt and fight [170].

Is intelligence really required? Plants can adapt to the environment; they process information slowly [170], but they cannot generate intelligent behavior, not to mention a civilization. Plants have neither specialized sensory organs nor a nervous system. They use hormones interpreted at the individual cell level to coordinate developmental decisions with resource allocation and growth.

Earth today is home to over seven hundred thousand species of insects. They are one of the most successful life forms on Earth, but they do not produce any intelligence.

Many multiverses have no technical civilizations. We inhabit a very unprovable universe, which is quite unusual.

6.5 SUMMA TECHNOLOGIAE

"Summa Technologiae" [170] is a 1964 book by the Polish author Stanisław Lem (Figure 6.7). The book tries to "examine the thorns of roses that have not flowered yet" to address future problems. Despite its age, the book has lost none of its relevance. We examine some of the problems described in the book in the context of our findings from the subsections about other civilizations.

The goal of scientific and technological development is to make humans independent of our environment by gaining mastery over it in the struggle for survival. Despite progress, humans remain helpless in the face of nature, for example, to earthquakes, to the rare danger of meteorites and to climate change. The only solution to overcoming these problems is to follow the path of science and technology. As they make scientific progress, humans hardly know what they are doing. For example, the destruction of life on earth through atomic war was not the initial goal of physics. Instead, technologies are born from accidents, such as, for example, the invention of porcelain or the discovery of penicillin. The accidents happen through the growth of

Figure 6.7 Stanisław Herman Lem (1921–2006), a Polish writer of science fiction with a toy cosmonaut.

knowledge. The evolution of present technologies is based on known knowledge: the first automobile looked like the coach; the first airplane, like the kite. A similar explanation is often given for the mind as that used for current technology. First it was like a hydraulic pump, then a mechanical clock, then a computer and now a quantum computer. At some point, a technology is abandoned in favor of a new one. As with the analogy, no one thinks today that our brain works like a mechanical device. We think that the brain is related to a computer without realizing that a mechanical device and a computer describe the same mechanism on different levels.

The relation "if A then B" is often based on magical thinking in which unknown knowledge is deduced. An explanation is often based on the false theoretical knowledge of the time followed by a chain of pseudo-explanations.

The process of inventions was initiated by abolishing Aristotelianism as a dogma and adopting empiricism as a directive for all cognitive activity [170]. We passed from rules to statistical laws, from rigid causality to probability and from simplicity to complexity. By abolishing the Copenhagen interpretation, the first quantum

Figure 6.8 IBM Q System with refrigeration technology built around a current quantum processor to ensure its calculations are accurate. The quantum processor has to sit inside a shield to protect itself from electromagnetic radiation. In 2021, IBM reached 127-quantum bits (qubits) with its new Eagle processor, its first quantum processor to contain more than 100 qubits.

computers were developed. We seem to have difficulty accepting the multiverse description of our reality. This difficulty will change as the first quantum computers develop, which will perform calculations on a scale unimaginable today, permitting computational capacity beyond our present capabilities (see Figure 6.8).

Are there any limits to scientific progress? Are we not limited by our human nature? Science has grown exponentially since the seventeenth century; however, exponential growth cannot continue indefinitely. No one knows how many scientific journals are in existence, but several estimates point to approximately 30,000, with close to two million articles published each year. The number of scientists is growing exponentially. Theoretical knowledge has progressed beyond the practical knowledge used in industry. The growth fluctuates of theoretical knowledge. For example, the Copenhagen interpretation stopped the invention of quantum computers, and computationalism stopped the search for the explanation of the mind and lead to the naïve assumption that we are simply machines comparable to robots. As Stanisław Lem states:

"The entire history of science demonstrates that technological progress is always a consequence of discoveries gained by means of pure research, which is not focused on any practical goals. The reverse process, in which new knowledge emerges from a

technology that is already in use, has in turn been extremely rare and hence seems quite unusual."

Society decides which areas of research are important, i.e., motivated by practical applications, and which to exclude. Thomas Watson, president of IBM, 1943 stated, "*I think there is a world market for maybe five computers.*" Today, many computer departments see quantum computing as being outside the realm of computer science. It is interpreted as a mythology without any practical future application. A massive concentration of experts in popular areas is promoted through social networks, which, like deep learning, slow scientific progress. Instead of searching for new machine-learning algorithms motivated by cognitive science, most AI researchers do not want to take risks in their career and spend time on minor improvements of existing deep-learning algorithms. A similar problem happened earlier with the AI programming languages Prolog and LISP, which hindered scientific progress by claiming to help us understand intelligence.

6.5.1 Artificial Intelligence

Artificial intelligence (AI) is haunted by the medieval myth of the homunculus, an artificially created intelligent being like a humanoid robot or an android with a mind of its own [170]. This assumption is based on an inaccurate presumption of computationalism. A computer can exhibit intelligent behavior without a consciousness. When we play computer chess and the computer is a much better player than we are, we do not assume that the corresponding program has developed any trace of consciousness. Fifty years ago, the assumption was that the computer playing chess demonstrated the roots of human intelligence that could be never reached due to the exponential demands of computational resources. No one thought that, in addition to computer resources, efficient heuristics could solve the problem. Today, computer chess programs are not seen as intelligent at all.

Is a human being superior to an artificial intelligence program? The question does not make any sense. A car is faster than we are, but we do not compete with cars. Instead, we use them as transportation tools. The same could be said about artificial intelligence: we do not compete with it, but we use it as a tool.

We could have an extremely large database that stores an answer to every question. A machine would use this database to answer our questions. Such a machine would clearly seem intelligent to us, but, knowing its working principle, we would never assume that it has a consciousness. Modern deep-learning algorithms work by the same principle. The only difference is that they compress the large database.

Consciousness or mind is not a technological problem because an engineer is not interested in whether a machine has consciousness but only in whether it works. Could consciousness emerge accidentally? We still do not understand how the brain works; it remains a kind of black box to us. Therefore, we naïvely assume that, if we do not understand a system due to its complexity, it will develop a mind, and consciousness will emerge accidentally.

6.5.2 Virtual Reality and Immortality

Interpreting information can be troublesome. A book is written for a reader, but if there is no reader, the book has no meaning. Instead of a book, we can create some virtual realities that simulate experiences that can be like or completely different from the real world. Currently, virtual reality systems use virtual reality headsets to generate realistic images on a small screen directly in front of the user's eyes and speakers to generate sounds that simulate a user's physical presence in a virtual environment.

Virtual reality headsets allow a person to look around the artificial world, move around in it, and interact with it. Some users may experience virtual reality sickness. This phenomenon occurs when a person's exposure to a virtual environment causes symptoms that are similar to motion sickness. Additionally, seizures or blackouts can occur while using virtual reality headsets. Even a simulated reality requires a human body to feel it. Virtual reality does not permit us to escape our bodies. We cannot simulate everything. Even if we connect the brain to a computer, the brain still perceives the splitting in the multiverse. Some realities can develop brain cancer, while others cannot.

Another romantic wish motivated by science is the desire for immortality, for example, digital immortality [205]. This assumption is based on an incorrect presumption of computationalism that does not consider the multiverse reality of the world. Imagine we could live forever in a computer simulation; we would not need to sleep. What would we do with our infinite life? We would probably get artificially drunk with the aim to forget ourselves, which would lead to a paradox of wanting to not exist and a digital suicide.

6.5.3 Limitations

Technological evolution and science will give us increasing abilities. However, science alone cannot answer some questions, such as those about morality and ethics, love, the meaning of our lives or the creation of beauty. A new revolution is underway, a technological one through quantum computers and a philosophical one that indicates to us our place in the world and that radically changes our picture of the world and our lives.

The explanation hypothesis and the multiverse are based on the many-worlds theory (multiverse) or many-worlds interpretation (MWI), and guarantee us free will. Consciousness senses the brain and the world; it is the unchangeable, nonphysical part of us. We are defined by our life choices and are fully responsible for them; this explanation corresponds to the philosophy of Existentialism.

Jean-Paul Charles Aymard Sartre (1905–1980) was a key figure in the philosophy of Existentialism. He argued that individuals shape themselves by existing; the actual life of the individuals is what constitutes life. Human beings, through their own consciousness, create their own values and determine a meaning for their lives. When we leave the world in which human life takes place and move into an abstract reality, only emptiness remains.

A person can choose to act in different ways and to be a good instead of a cruel person. Nothing fixes our purpose but us; our projects have no weight or inertia except for our endorsement of them. As Sartre said in his lecture, Existentialism is Humanism: "*Man first of all exists, encounters himself, surges up in the world and defines himself afterwards.*"

A renaissance of humanities and philosophy will occur in addition to science. Science will understand that it cannot explain ethics or the meaning of life. It can, however, help us to understand the nature of our world while indicating its limits, permitting free space for spirituality. Science will find that we are not merely machines and that neither love nor ethics can be proven rationally.

Bibliography

[1] S. Aaronson. Quantum machine learning algorithms: Read the fine print. *Nature Physics*, 11:291–293, 2015.

[2] J.S. Aikins. Prototypical knowledge for expert systems. *Artificial Intelligence*, 20:163–210, 1986.

[3] E Aïmeur, B. Brassard, and S Gambs. Quantum speed-up for unsupervised learning. *Machine Learning*, 90:261–287, 2013.

[4] Igor Aizenberg, Naum N. Aizenberg, and Joos P.L. Vandewalle. *Multi-Valued and Universal Binary Neurons*. Springer, 2000.

[5] David Albert and Barry Loewer. Interpreting the many-worlds interpretation. *Synthese*, 77:195–213, 1988.

[6] L. Amico, R. Fazio, A. Osterloh, and V. Vedral. Entanglement in many-body systems. *Reviews of Modern Physics*, 80(2):517–576, 2008.

[7] John R. Anderson. *Cognitive Psychology and its Implications*. W. H. Freeman and Company, fourth edition, 1995.

[8] Joseph Anderson and Barbara Anderson. The myth of persistence of vision revisited. *Journal of Film and Video*, 45(1):3–12, 1993.

[9] J.R. Anderson. *The Architecture of Cognition*. Harvard University Press, 1983.

[10] Aristotle. *De Anima (On the Soul) (Penguin Classics)*. Penguin Classics, 1987.

[11] Aristotle. *On Sense And The Sensible*. Kessinger Publishing, LLC, 2010.

[12] Aristotle. *On Dreams*. CreateSpace Independent Publishing Platform, 2016.

[13] Aristotle. *Aristotle On Fallacies: Or, the Sophistici Elenchi*. Franklin Classics, 2018.

[14] E. Aserinsky and N. Kleitman. Regularly occurring periods of eye motility, and concomitant phenomena, during sleep. *Science*, 118(3062):273–274, 1953.

[15] E. Aserinsky and N. Kleitman. Two types of ocular motility occurring in sleep. *Journal of Applied Physiology*, 8(1):1–10, 1955.

[16] W. Aspray. *John von Neumann and the origins of modern computing*. History of computing. MIT Press, 1990.

[17] Bernard Baars. *A Cognitive Theory of Consciousness.* Cambridge University Press, 1988.

[18] Bernard Baars. *In the Theater of Consciousness: The Workspace of the Mind.* Oxford University Press, 1997.

[19] Francis Bacon. *Francis Bacon: The New Organon (Cambridge Texts in the History of Philosophy).* Cambridge University Press, 2000.

[20] Dana H. Ballard. *An Introduction to Natural Computation.* The MIT Press, 1997.

[21] Y. Bao, E. Pöppel, W. Liang, and T. Yang. When is the right time? a little later! – delayed responses show better temporal control. *Procedia - Social and Behavioral Sciences*, 126:199–200, 2014.

[22] Jeffrey A. Barrett and Peter Byrne. *The Everett Interpretation of Quantum Mechanics, Collected Works 1955-1980 with Commentary.* Princeton University Press, 2012.

[23] Zygmunt Bauman. *Modernity and the Holocaust.* Cornell University Press, 2006.

[24] Friedrich Beck and John C. Eccles. Quantum aspects of brain activity and the role of consciousness. *Proceedings of the National Academy of Science Usa*, 89:11357–11361, 1992.

[25] Paul Benioff. Quantum robots and environments. *ArXiv Quantum Physics e-prints*, 1998.

[26] C.H. Bennett. Logical reversibility of computation. *IBM Journal of Research and Development*, 17:525–532, November 1973.

[27] Charles H. Bennett. The thermodynamics of computation–a review. *International Journal of Theoretical Physics*, 21(12):905–940, 1982.

[28] Charles H. Bennett. Notes on landauer's principle, reversible computation, and maxwell's demon. *Studies in History and Philosophy of Science Part B: Studies in History and Philosophy of Modern Physics*, 34(3):501–510, 2003.

[29] Charles H. Bennett, Ethan Bernstein, Gilles Brassard, and Umesh Vazirani. Strengths and weaknesses of quantum computing. ArXiv Quantum Physics e-prints, 1997.

[30] Charles H. Bennett, Gilles Brassard, Claude Crépeau, Richard Jozsa, Asher Peres, and William K. Wootters. Teleporting an unknown quantum state via dual classical and einstein-podolsky-rosen channels. *Phys. Rev. Lett.*, 70(13):1895–1899, Mar 1993.

[31] Elwyn R. Berlekamp, John H. Conway, and Richard K. Guy. *Winning Ways for Your Mathematical Plays, Vol. 2 2nd Edition.* A K Peters/CRC Press, 2003.

[32] C. Bernard. *An Introduction to the Study of Experimental Medicine.* New York, NY: Dover, 1957.

[33] Shmuel Bialy and Abraham Loeb. Could solar radiation pressure explain 'oumuamua's peculiar acceleration? *The Astrophysical Journal Letters*, 868(1):1–5, 2018.

[34] James Binney and David Skinner. *The Physics of Quantum Mechanics.* Oxford University Press, 2014.

[35] Susan Blackmore. *Consciousness: A Very Short Introduction.* Oxford University Press, 2005.

[36] Chris Bleakley. *Poems That Solve Puzzles: The History and Science of Algorithms.* Oxford University Press, 2020.

[37] Raphael Bousso and Leonard Susskind. Multiverse interpretation of quantum mechanics. *Physical Reviewd D*, 85(4):045007, 2012.

[38] Carl B. Boyer and Uta C. Merzbach. *A History of Mathematics 3rd Edition.* Wiley, 2010.

[39] Michel Boyer, Gilles Brassard, Peter Hoeyer, and Alain Tapp. Tight bounds on quantum searching. *Fortschritte der Physik*, 46:493, 1998.

[40] Valentino Braitenberg. *Gehirngespinste, Neuroanatomie für kybernetisch Interessierte.* Springer-Verlag, 1973.

[41] Valentino Braitenberg. Cell assemblies in the cerebral cortex. In R. Heim and G.Palm, editors, *Theoretical Approaches to Complex Systems*, pages 171–188. Springer-Verlag, 1978.

[42] Valentino Braitenberg. *Vehicles: Experiments in Synthetic Psychology.* The MIT Press, 1984.

[43] J Brooke, D. Bitko, T.F. Rosenbaum, and G. Aeppli. Quantum annealing of a disordered magnet. *Science*, 284(5415):779–781, 1999.

[44] Christopher R. Browning. *Fateful Months: Essays on the Emergence of the Final Solution.* Holmes and Meier Publishers, 1991.

[45] L. Brownston, R. Farell, E. Kant, and N. Martin. *Programming Expert Systems in OPS5: An Introduction to Rule-Based Programming.* Addison-Wesley, 1985.

[46] S. Brunak and B. Lautrup. *Neural Networks Computers with Intuition.* World Scientific, 1990.

[47] Arthur E. Bryson and Yu-Ch Ho. *Applied Optimal Control : Optimization, Estimation, and Control.* Blaisdell Pub. Co., 1969.

[48] Jerome Busemeyer and Zheng Wang. Quantum cognition: Key issues and discussion. *Topics in Cognitive Science*, 6:43–46, 2014.

[49] Jerome Busemeyer, Zheng Wang, and Jennifer Trueblood. Hierarchical bayesian estimation of quantum decision model parameters. In *Proceedings of the 6th International Symposium on Quantum Interactions*, pages 80–89, 2012.

[50] Jerome R. Busemeyer and Jennifer Trueblood. Comparison of quantum and bayesian inference models. In Peter Bruza, Donald Sofge, William Lawless, Keith van Rijsbergen, and Matthias Klusch, editors, *Quantum Interaction*, volume 5494 of *Lecture Notes in Computer Science*, pages 29–43. Springer, 2009.

[51] Jerome R. Busemeyer, Zheng Wang, and Ariane Lambert-Mogiliansky. Empirical comparison of markov and quantum models of decision making. *Journal of Mathematical Psychology*, 53(5):423–433, 2009.

[52] Jerome R. Busemeyer, Zheng Wang, and James T. Townsend. Quantum dynamics of human decision-making. *Journal of Mathematical Psychology*, 50(3):220–241, 2006.

[53] Peter Byrne. The many worlds of hugh everett. *Scientific American Magazine*, pages 98–105, December 2007.

[54] Rpberto Cabeza and Alan Kingstone. *Hanbook of Functional Neuroimaging of Cognition.* The MIT Press, 2001.

[55] David J. Chalmers. *The Conscious Mind: In Search of a Fundamental Theory.* Oxford University Press, 1996.

[56] Anne Charlton. Medicinal uses of tobacco in history. *Journal of the Royal Society of Medicine*, 97(6):292–296, 2004.

[57] Alonzo Church. A note on the entscheidungsproblem. *Journal of Symbolic Logic*, 1(1):40–41, March 1936.

[58] Alonzo Church. An unsolvable problem of elementary number theory. *American Journal of Mathematics*, 58(2):345–363, April 1936.

[59] Alonzo Church. *The Calculi of Lambda-Conversion.* Annals of Mathematics Studies. Princeton University Press, Princeton, New Jersey, USA, 1941.

[60] Patricia S. Churchland and Terrence J. Sejnowski. *The Computational Brain.* The MIT Press, 1994.

[61] Paul M. Churchland. *Matter and Consciousness: A Contemporary Introduction to the Philosophy of Mind.* The MIT Press, 1988.

[62] Paul M. Churchland. *The Engine of Reason, the Seat of the Soul*. The MIT Press, 1995.

[63] Michael A. Cohen and Daniel C. Dennett. Consciousness cannot be separated from function. *Trends in Cognitive Science*, 15(8):358–364, 2011.

[64] Dennis Coon and John O. Mitterers. *Introduction to Psychology Gateways to Mind and Behavior, 13th Edition*. Wadsworth Publishing, 2012.

[65] Nicolaus Copernicus. *On the Revolutions of Heavenly Spheres (Great Minds Series)*. Prometheus, 1995.

[66] T. H. Cormen, C. E. Leiserson, L. R. Rivest, and C. Stein. *Introduction to Algorithms*. Second edition. MIT Press, 2001.

[67] Thomas H. Cormen, Charles E. Leiserson, Ronald L. Rivest, and Clifford Stein. *Introduction to Algorithms, 2/e*. MIT Press, 2001.

[68] Joao Crespo and Andreas Wichert. Reinforcement learning applied to games. *SN Applied Sciences*, 2:1–16, 2020.

[69] Francis Crick. Function of the thalmic reticular complex: The searchligth hypothesis. In B. J. Baars, W. P. Banks, and J.B. Newman, editors, *Essential Sources in the Scientific Study of Consciousness*. MIT Press, 2003.

[70] Francis Crick and Graeme Mitchison. The function of dream sleep. *Nature*, 304:111–114, 1983.

[71] Christopher Cullen. *Astronomy and Mathematics in Ancient China*. Cambridge University Press, 1996.

[72] Gregory Curtis. *The Cave Painters: Probing the Mysteries of the World's First Artists*. Alfred A. Knopf, 2006.

[73] Guthrie R. Dale. *The Nature of Prehistoric Art*. University of Chicago Press, 2006.

[74] Antonio Damasio. *The Feeling of What Happens: Body and Emotion in the Making of Consciousness*. Mariner Books, 2000.

[75] Antonio Damasio. *Feeling and Knowing: Making Minds Conscious*. Pantheon, 2021.

[76] Peter Dayan, Geoffrey E. Hinton, and Radford M. Neal. The helmholtz machine. *Neural Computation*, 7:889–904, 1995.

[77] Rina Dechter. Learning while searching in constraint-satisfaction problems. In *AAAI-86 Proceedings*, pages 178–183, 1986.

[78] W. Dement and N. Kleitman. The relation of eye movements during sleep to dream activity: an objective method for the study of dreaming. *Journal of experimental psychology*, 53(5):339, 1957.

[79] Daniel C. Dennett. *Consciousness Explained*. Back Bay Books, 1992.

[80] Rene Descartes. *Meditations and Other Metaphysical Writings (Penguin Classics)*. Penguin Classics, 1999.

[81] David Deutsch. Quantum theory, the church-turing principle and the universal quantum computer. In *Proceedings of the Royal Society of London- Series A, Mathematical and Physical Sciences*, volume 400, pages 97–117, 1985.

[82] David Deutsch. "many minds" interpretations of quantum mechanics. *British Journal for the Philosophy of Science*, 47:222–228, 1996.

[83] David Deutsch. *The Fabric of Reality*. Penguin Group, 1997.

[84] David Deutsch. Quantum theory of probability and decisions. *Proceedings of the Royal Society A*, 455:3129–97, 1999.

[85] David Deutsch. *The Beginning of Infinity: Explanations That Transform the World*. Penguin Books, 2012.

[86] David Deutsch. Constructor theory. *Synthese*, 190(18), 2013.

[87] L. DiCarlo, M. D. Reed, L. Sun, B. R. Johnson, J. M. Chow, J. M. Gambetta, L. Frunzio, S. M. Girvin, M. H. Devoret, and R. J. Schoelkopf. Preparation and measurement of three-qubit entanglement in a superconducting circuit. *Nature*, 467(7315):574–578, 09 2010.

[88] C.J. Downing and S. Oinker. The spatial structure of visual attention. In M.I. Posner and O.S.M. Marin, editors, *Attention and performance XI*, pages 171–187. Erlbaum, Hillsdale, N.J., 1985.

[89] Richard O. Duda, Peter E. Hart, and David G. Stork. *Pattern Classification (2nd Edition)*. Wiley-Interscience, 2000.

[90] Werner Ebeling. *Chaos-Ordnung Information*. Verlag Harri Deutsch, 1989.

[91] John C. Eccles. *How the SELF Controls Its BRAIN*. Springer, 1994.

[92] Chris Eliasmith and Charles H. Anderson. *Neural Engineering: Computation, Representation, and Dynamics in Neurobiological Systems*. The MIT Press, 2004.

[93] Larry G. Epstein. A definition of uncertainty aversion. *The Review of Economic Studies*, 66(3):579–608, 1999.

[94] G.W. Ernst and A. Newell. *GPS: A Case Study in Generality and Problem Solving*. Academic Press, 1969.

[95] Euclid. *Euclid's Elements*. Green Lion Press;, 2002.

[96] Hugh Everett. "relative state" formulation of quantum mechanics. *Reviews of Modern Physics*, 29(3):454–462, 1957.

[97] Michael Feirtag and Walle J Nauta. *Neuroanatomie*. Spektrum Akademischer Verlag, 1990.

[98] David Ferrucci, Anthony Levas, Sugato Bagchi, David Gondek, and Erik T. Mueller. Watson: Beyond jeopardy! *Artificial Intelligence*, 199:93–105, 2013.

[99] Maurice A. Finocchiaro. *The Trial of Galileo: Essential Documents (Hackett Classics)*. Hackett Publishing Company, 2014.

[100] Dario Floreano and Claudio Mattiussi. *Bio-Inspired Artificial Intelligence: Theories, Methods, and Technologies*. The MIT Press, 2008.

[101] Sigmund Freud. *The Basic Writings of Sigmund Freud*. Modern Library, 1995.

[102] Sigmund Freud. *The Interpretation of Dreams: The Complete and Definitive Text*. Basic Books, 2010.

[103] J. P. Frisby and J. V. Stone. *Seeing, The Computational Approach to Biological Vision, 2nd Edition*. MIT Press, 2010.

[104] Jean-Gabriel Ganascia. La conception des systemes experts. *La Recherche*, Octobre 1985.

[105] Martin Gardner. Mathematical games: Mathematical games: The random number omega bids fair to hold the mzsteries of the universe. *Scientific American*, pages 20–30, 1979.

[106] Richard J. Gaylord and Paul R. Wellin. *Computer Simulations with Mathematica*. Spriner Verlag, 1995.

[107] Danko D. Georgiev. *Quantum Information and Consciousness: A Gentle Introduction*. CRC Press, 2018.

[108] S. Ghosh, T. F. Rosenbaum, G. Aeppli, and S. N. Coppersmith. Entangled quantum state of magnetic dipoles. *Nature*, 425:48–51, 2003.

[109] James Gleick. *Chaos: Making a New Science*. Penguin Books, 2008.

[110] Mark A. Gluck, Eduardo Mercado, and Catherine E. Myers. *Learning and Memory: From Brain to Behavior, Fourth edition*. Worth Publishers, 2019.

[111] Kurt Gödel. *On Formally Undecidable Propositions of Principia Mathematica and Related Systems*. Dover Publications, 1992.

[112] Peter Godfrey-Smith. *Other Minds: The Octopus, the Sea, and the Deep Origins of Consciousness*. Farrar, Straus and Giroux, 2016.

[113] Hilary Greaves. Probability in the everett interpretation. *Philosophy Compass*, 1(109-128), 2007.

[114] Hilary Greaves. Understanding deutsch's probability in a deterministic multiverse. *Studies in History and Philosophy of Modern Physics*, 3(423-456), 35.

[115] C. G. Gross. Claude bernard and the constancy of the internal environment. *Neuroscientist*, 4:380–385, 1998.

[116] C.G. Gross and Mishkin. The neural basis of stimulus equivalence across retinal translation. In S. Harnad, R. Dorty, J. Jaynes, L. Goldstein, and Krauthamer, editors, *Lateralization in the nervous system*, pages 109–122. Academic Press, New York, 1977.

[117] Lov K. Grover. A fast quantum mechanical algorithm for database search. In *STOC '96: Proceedings of the Twenty-Eighth Annual ACM Symposium on Theory of Computing*, pages 212–219, New York, NY, USA, 1996. ACM.

[118] Lov K. Grover. Quantum mechanics helps in searching for a needle in a haystack. *Physical Review Letters*, 79:325, 1997.

[119] Lov K. Grover. A framework for fast quantum mechanical algorithms. In *STOC '98: Proceedings of the Thirtieth Annual ACM Symposium on Theory of Computing*, pages 53–62, New York, NY, USA, 1998. ACM.

[120] Lov K. Grover. Quantum computers can search rapidly by using almost any transformation. *Phys. Rev. Lett.*, 80(19):4329–4332, May 1998.

[121] David J. Gunkel. *Robot Rights*. The MIT Press, 2018.

[122] Tim Haines. *Walking with Dinosaurs - a Natural History*. BBC Worldwide Limited, 1999.

[123] Herman Haken. *Synergetik*. Springer-Verlag, 1990.

[124] Hermann Haken. *Synergetic Computers and Cognition*. Springer-Verlag, 1991.

[125] Stuart Hameroff and Roger Penrose. Orchestrated reduction of quantum coherencein brain microtubules: A model for consciousnes. *Mathematics and Computersin Simulation*, 40:453–480, 1996.

[126] Stuart Hameroff and RogerPenrose. Consciousness in the universe: A review of the 'orch or' theory. *Physics of Life Reviews*, 11(1):39–78, 2014.

[127] Yannis Hamilakis. Food technologies/technologies of the body: The social context of wine and oil production and consumption in bronze age crete. *World Archaeology*, 31(1):38–54, 1999.

[128] K. Hammond. *Case-Based Planning: Viewing Planning as a Memory Task*. Academic Press, New York, 1989.

[129] A. Harrow, A. Hassidim, and S. Lloyd. Quantum algorithm for solving linear systems of equations. *Physical Review Letters*, 103:150502, 2009.

[130] D. Hebb. *The Organization of Behavior*. John Wiley, New York, 1949.

[131] D. Hebb. *Textbook of Psychology*. Sanders, Philadelphia London Toronto, 1958.

[132] Werner Heisenberg. *The Physical Principles of the Quantum Theory*. Courier Dover Publications, 1949.

[133] John Hertz, Anders Krogh, and Richard G. Palmer. *Introduction to the Theory of Neural Computation*. Addison-Wesley, 1991.

[134] Anthony Hey and Patrick Walters. *The New Quantum Universe*. Cambridge University Press, 2003.

[135] Douglas Hofstadter. *I Am a Strange Loop*. Basic Books, 2007.

[136] J. J. Hopfield. Neural networks and physical systems with emergent collective computational abilities. *Proceedings of the National Academy of Sciences of the USA*, 79(8):2554–2558, 1982.

[137] D. H. Hubel. *Eye, Brain, and Vision*. Scientific Ammerican Library, Oxford, England, 1988.

[138] D. H. Hubel and T. N. Wiesel. Receptive fields, binocular interaction and functional architecture in the cat's visual cortex. *J Physiol*, 160:106–154, Jan 1962.

[139] D. H. Hubel and T. N. Wiesel. Receptive fields and functional architecture of monkey striate cortex. *J Physiol*, 195(1):215–243, Mar 1968.

[140] Anthony Hyman. *Charles Babbage: Pioneer of the Computer*. Princeton University Press, 1985.

[141] Peter Jackson. *Introduction to Expert Systems*. Addison-Wesley, third edition, 1999.

[142] Robert G. Jahn and Brenda J. Dunne. *Margins of Reality: The Role of Consciousness in the Physical World*. A Heverst Book, 1988.

[143] Karl Jaspers. *Philosophy of Existence*. University of Pennsylvania Press, 1971.

[144] M. W. Johnson, M. H. S. Amin, S. Gildert, T. Lanting, F. Hamze, N. Dickson, R. Harris, A. J. Berkley, J. Johansson, P. Bunyk, E. M. Chapple, C. Enderud, J. P. Hilton, K. Karimi, E. Ladizinsky, N. Ladizinsky, T. Oh, I. Perminov, C. Rich, M. C. Thom, E. Tolkacheva, C. J. S. Truncik, S. Uchaikin, J. Wang, B. Wilson, and G. Rose. Quantum annealing with manufactured spins. *Nature*, 473(7346):194–198, 05 2011.

[145] G. Kahn, A. Kepner, and J. Pepper. Test: a model-driven application shell. In *National Conference on Artificial Intelligence*, pages 814–18, 1987.

[146] K. Karimi, N. G. Dickson, F. Hamze, M. H. S. Amin, M. Drew-Brook, F. A. Chudak, P. I. Bunyk, W. G. Macready, and G. Rose. Investigating the Performance of an Adiabatic Quantum Optimization Processor. *ArXiv e-prints*, June 2010.

[147] Philip R. Kaye, Raymond Laflamme, and Michele Mosca. *An Introduction to Quantum Computing*. Oxford University Press, USA, 2007.

[148] Julian Keenan, Gordon G. Gallup, and Dean Falk. *The Face in the Mirror: The Search for the Origins of Consciousness*. Ecco, 2003.

[149] Andrei Khrennikov. Quantum-like model of cognitive decision making and information processing. *Journal of BioSystems*, 95:179–187, 2009.

[150] Alexei Kitaev. Quantum measurements and the abelian stabilizer problem. *Electronic Colloquium on Computational Complexity*, 3(TR96-003), 1996.

[151] P. Klahr and D.A. Waterman. *Expert Systems: Techniques, Tools and Applications*. Addison-Wesley, 1986.

[152] Stanley B. Klein and Shaun Nichols. Memory and the sense of personal identity. *Mind*, 121(677–702), 2012.

[153] Hüseyin Kocak. *Differential and Difference Equations through Computer Experiments*. Springer, 1989.

[154] Christof Koch. *The Quest for Consciousness: A Neurobiological Approach*. Roberts and Co, 2004.

[155] Christof Koch. *Consciousness: Confessions of a Romantic Reductionist*. The MIT Press, 2012.

[156] Teuvo Kohonen. *Self-Organization and Associative Memory*. Springer-Verlag, 3 edition, 1989.

[157] Richard E. Korf. Depth-first iterative-deepening : An optimal admissible tree search. *Artificial Intelligence*, 27(1):97 – 109, 1985.

[158] S. M. Kosslyn. If neuroimaging is the answer, what is the question? *Philosophical Transactions of the Royal Society of London B Biological Sciences*, 354:1283–1294, 1999.

[159] Stephen M. Kosslyn. *Image and Brain, The Resolution of the Imagery Debate*. The MIT Press, 1994.

[160] Raymond Kurzweil. *The Age of Intelligent Machines*. The MIT Press, 1990.

[161] Kushyar Ibn Labban. *Principles of Hindu Reckoning.* University of Wisconsin Press, 1965.

[162] R.C. Lacher. Expert networks: Paradigmatic conflict, technological rapprochement. *Minds and Machines*, 3:53–71, 1993.

[163] David Lambert. *Collins Guide to Dinosaurs.* Diagram Visual Information Ltd, 1983.

[164] David Lambert. *The Ultimate Dinosaur Book.* Dorling Kindersley, 1993.

[165] R. Landauer. Information is physical. In *Physics and Computation, 1992. PhysComp '92., Workshop on Physics and Computation*, pages 1–4, Oct 1992.

[166] Rolph Landauer. Irreversibility and heat generation in the computing process. *IBM Journal of Research and Development*, 5:183–191, 1961.

[167] Norman J. Lass and Charles M. Woodford. *Hearing Science Fundamentals.* Mosby, 2007.

[168] Simon B. Laughlin, Rob R. de Ruyter van Steveninck, and John C. Anderson. The metabolic cost of neural information. *Nature Neuroscience*, 1(1):36–41, 1998.

[169] Yann LeCun and Yoshua Bengio. *Convolutional networks for images, speech, and time series*, pages 255–258. MIT Press, Cambridge, MA, USA, 1998.

[170] Stanisław Lem. *Summa Technologiae.* Univ Of Minnesota Press, 2014.

[171] Harry R. Lewis and Christos H. Papadimitriou. *Elements of the Theory of Computation.* Prentice Hall PTR, Upper Saddle River, NJ, USA, 1981.

[172] P. Lewis. Probability in everettian quantum mechanics. *Manuscrito*, 33:285–306, 2010.

[173] Benjamin Libet. *Mind Time - The Temporal Factor in Consciousness.* Harvard University Press, 2004.

[174] F. London and E. Bauer. The theory of observation in quantum mechanics. In J.A. Wheeler and W.H. Zurek, editors, *Quantum Theory and Measurement,.* Princeton University Press, 1983.

[175] Peter Lucas and Linda van der Gaag. *Principles of Expert Systems.* Addison-Wesley, 1991.

[176] George F. Luger and William A. Stubblefield. *Artificial Intelligence, Structures and Strategies for Complex Problem Solving.* Addison-Wesley, third edition, 1998.

[177] Jean-François Lyotard. *La Condition Postmoderne.* Les Editions de Minuit, 1979.

[178] Jean-François Lyotard. *Le Différend.* Les Editions de Minuit, 1983.

[179] Malcolm Macmillan. *An Odd Kind of Fame: Stories of Phineas Gage.* A Bradford Book, 2002.

[180] Benoit B Mandelbrot. *The Fractal Geometry of Nature.* W. H. Freeman and Co., 1983.

[181] Benoit B Mandelbrot. *The Fractal Geometry of Nature.* Echo Point Books and Media, LLC, 2021.

[182] R.A. Marcus and Sutin N. Electron transfers in chemistry and biology. *Biochimica et Biophysica Acta,* 811(3):256, 1985.

[183] Cordier Marie-Odilie. Le systemes experts. *La Recherche,* Janvier 1984.

[184] John Martin. *Neuroanatomy Text and Atlas, Fifth Edition.* McGraw Hill, 2020.

[185] James Clerk Maxwell. *Theory of Heat (9ed).* Courier Dover Publications, 2001.

[186] J.L. McClelland and D.E. Rumelhart. *Explorations in Parallel Distributed Processing - IBM version.* The MIT Press, 1986.

[187] J.L. McClelland and D.E. Rumelhart. *Explorations in the Microstructure of Cognition. Volume 1: Foundations.* The MIT Press, 1986.

[188] J.L. McClelland and D.E. Rumelhart. *Explorations in the Microstructure of Cognition. Volume 2: Psychological and Biological Models.* The MIT Press, 1986.

[189] Patrick McGovern, Mindia Jalabadzeand Stephen Batiuk, Michael P. Callahan, Karen E. Smith, Gretchen R. Hall, Eliso Kvavadze, David Maghradze, Nana Rusishvili, Laurent Bouby, Osvaldo Failla, Gabriele Cola, Luigi Mariani, Elisabetta Boaretto, Roberto Bacilieri, Patrice This, Nathan Wales, and David Lordkipanidze. Early neolithic wine of georgia in the south caucasus. *PNAS,* 114(48), 2017.

[190] J. C. McKeown. *A Cabinet of Greek Curiosities: Strange Tales and Surprising Facts from the Cradle of Western Civilization.* Oxford University Press, 2013.

[191] Thomas Metzinger. *Being No One. The Self-Model Theory of Subjectivity.* MIT Press, 2003.

[192] Thomas Metzinger. Artificial suffering: An argument for a global moratorium on synthetic phenomenology. *Journal of Artificial Intelligence and Consciousness,* 8(1):43–66, 2021.

[193] Marvin Minsky. *The Society of Mind.* Simon and Schuster, New York, 1986.

[194] Marvin Minsky and Seymour Papert. *Perceptrons: An Introduction to Computational Geometry.* MIT Press, 1972.

[195] Graeme Mitchison and Richard Jozsa. Counterfactual computation. *Proceedings of the Royal Society of London A*, 457:1175–1193, 2009.

[196] A. Murschel. The structure and function of ptolemy's physical hypotheses of planetary motion. *Journal for the History of Astronomy*, 26(1):33–61, 1995.

[197] Thomas Nagel. *Mind and Cosmos: Why the Materialist Neo-Darwinian Conception of Nature is Almost Certainly False*. Oxford University Press, 2012.

[198] O. Neugebauer. *A History of Ancient Mathematical Astronomy. Studies in the History of Mathematics and Physical Sciences*. Springer-Verlag, 1975.

[199] M. Newborn. *Kasparov versus Deep Blue, Computer Chess Comes of Age*. Springer-Verlag, 1997.

[200] A. Newell and H. Simon. Computer science as empirical inquiry: symbols and search. *Communication of the ACM*, 19(3):113–126, 1976.

[201] A. Newell and H.A. Simon. *Human Problem Solving*. Prentice-Hall, 1972.

[202] Allen Newell. *Unified Theories of Cognition*. Harvard University Press, 1990.

[203] Sir Isaac Newton. *The Principia: The Authoritative Translation and Guide: Mathematical Principles of Natural Philosophy*. University of California Press, 2016.

[204] Nils J. Nilsson. *Principles of Artificial Intelligence*. Springer-Verlag, 1982.

[205] Arlindo Oliveira. *The Digital Mind: How Science Is Redefining Humanity*. The MIT Press, 2017.

[206] Günther Palm. *Neural Assemblies, an Alternative Approach to Artificial Intelligence*. Springer-Verlag, 1982.

[207] Günther Palm. Cell assemblies, coherence, and corticohippocampal interplay. *Hippocampus*, 3:219–226, 1993.

[208] John Palmer. *Parmenides and Presocratic Philosophy*. Oxford University Press;, 2010.

[209] Jian Pan, Yudong Cao, Xiwei Yao, Zhaokai Li, Chenyong Ju, Xinhua Peng, Sabre Kais, Jiangfeng Du, and Jiangfeng Du. Experimental realization of quantum algorithm for solving linear systems of equations. *Physical Review A.*, 89(2):022313, 2014.

[210] D. B. Parker. Learning-logic: Casting the cortex of the human brain in silicon. Technical Report Tr-47, Center for Computational Research in Economics and Management Science. MIT Cambridge, MA, 1985.

[211] Jaana Parviainen and Mark Coeckelbergh. The political choreography of the sophia robot: beyond robot rights and citizenship to political performances for the social robotics market. *AI and SOCIETY*, 36:715–724, 2021.

[212] Roger Penrose. *The Emperor's New Mind: Concerning Computers, Minds, and the Laws of Physics*. Oxford University, 1989.

[213] Roger Penrose. *Shadows of the Mind: A Search for the Missing Science of Consciousness*. Oxford University Press, 1994.

[214] Roger Penrose. *Fashion, Faith, and Fantasy in the New Physics of the Universe*. Princeton University Press, 2016.

[215] Jeff Pepper. An expert system for automotive diagnosis. In Raymond Kurzweil, editor, *The Age of Intelligent Machines*, pages 330–335. MIT Press, 1990.

[216] Plato. *The Republic (Penguin Classics)*. Penguin Classics, 2007.

[217] Plato. *Timaeus and Critias (Penguin Classics)*. Penguin Classics, 2008.

[218] Henri Poincaré. *The Foundations of Science: Science and Hypothesis, the Value of Science, Science and Method*. Arkose Press, 2015.

[219] E. Pöppel, K. Schill, and N. von Steinbüchel. Sensory integration within temporally neutral system states: a hypothesis. *Naturwissenschaftem*, 77:89–91, 1990.

[220] Ernst Pöppel. Pre-semantically defined temporal windows for cognitive processing. *Philos. Trans. R. Soc. Lond. B Biol. Sci.*, 364(1525): 1887–1896. 2009.

[221] Ernst Pöppel. Perceptual identity and personal self: neurobiological reflections. In M. Fajkowska and M. M. Eysenck, editors, *Personality From Biological, Cognitive, and Social Perspectives*, pages 75–82. Clinton Corners, NY: Eliot Werner Public, 2010.

[222] Karl Popper. *Conjectures and Refutations: The Growth of Scientific Knowledge*. Routledge, 2002.

[223] Michael I. Posner and Marcus E. Raichle. *Images of Mind*. Scientific American Library, New York, 1994.

[224] John Preston and Mark Bishop. *Views into the Chinese Room: New Essays on Searle and Artificial Intelligence*. Clarendon Press, 2002.

[225] Ross Quillian. Semantic memory. In Marvin Minsky, editor, *Semantic Information Processing*, pages 227–270. MIT Press, 1968.

[226] Howard L. Resnikoff. *The Illusion of Reality*. Springer-Verlag, 1989.

[227] Eleanor Rieffel and Wolfgang Polak. *Quantum Computing - A Gentle Introduction*. The MIT Press, 2011.

[228] M. Riesenhuber and T. Poggio. Neural mechanisms of object recognition. *Current Opinion in Neurobiology*, 12:162–168, 2002.

[229] M. Riesenhuber, T. Poggio. Hierarchical models of object recognition in cortex. *Nature Neuroscience*, 2:1019–1025, 1999.

[230] Maximilian Riesenhuber and Tomaso Poggio. Models of object recognition. *Nat Neuroscience*, 3:1199–1204, 2000.

[231] Andrew Robinson. *Writing and Script: A Very Short Introduction*. Oxford University Press, 2000.

[232] Frank Rosenblatt. *Principles of neurodynamics: Perceptrons and the theory of brain mechanisms*. Spartan Books, 1962.

[233] Richard Rudgley. *The Lost Civilizations of the Stone Age*. Simon & Schuster., 2000. ·

[234] D. Russell and R. Sequin. Reconstruction of the small cretaceous theropod stenonychosaurus inequalis and a hypothetical dinosauroid. *Syllogeous*, 37:1–43, 1982.

[235] S.J. Russell and P. Norvig. *Artificial Intelligence: A Modern Approach*. Prentice Hall series in artificial intelligence. Prentice Hall, 2010.

[236] Stuart J. Russell and Peter Norvig. *Artificial Intelligemce: A Modern Approach*. Prentice-Hall, 1995.

[237] Stuart J. Russell and Peter Norvig. *Artificial Intelligemce: A Modern Approach*. Prentice-Hall, second edition, 2003.

[238] Gilbert Ryle. *The Concept of Mind*. Penguin Classics, 2000.

[239] Carl Sagan and I. S. Shklovskii. *Intelligent Life in the Universe*. Holden Day, 1984.

[240] Emik Samkian. VV-XPS, ein Expertensystem zur Ermittlung absetzbarer Werbungskosten und Sonderausgaben bei Einküften aus Vermitung und Verpachtung. Master's thesis, Universität des Saarlandes, Saarbrücken, Germany, 1992.

[241] N. K. Sandars. *The Epic of Gilgamesh (Penguin Epics)*. 2006, Penguin.

[242] Elizabeth Schechter. *Self-Consciousness and 'Split' Brains: The Minds' I*. Oxford University Press, 2018.

[243] Erwin Schrödinger. Die gegenwärtige situation in der quantenmechanik. *Naturwissenschaften*, 23(807), 1935.

[244] Maria Schuld and Nathan Killoran. Quantum machine learning in feature hilbert spaces. *Physical Review Letters*, 122, 2019.

[245] John Searle. Minds, brains and programs. *Behavioral and Brain Sciences*, 3:417–457, 1980.

[246] Miguel Angel Sebastián. Dreams: an empirical way to settle the discussion between cognitive and non-cognitive theories of consciousness. *Synthese*, 191:263–285, 2013.

[247] R. Serra and G. Zanarini. *Complex Systems and Cognitive Processes*. Springer-Verlag, 1990.

[248] Thomas Serre. Deep learning: The good, the bad, and the ugly. *Annual Review of Vision Science*, 5:399–426, 2019.

[249] C. E. Shannon. Computers and automata. *Proceedings of the I.R.E.*, 41:1253–1241, 1953.

[250] Claude E. Shannon. A mathematical theory of communication. *Bell System Technical Journal*, pages 1–54, 1948.

[251] L. Shastri. *Semantic Networks: An Evidential Formulation and its Connectionistic Realization*. Morgan Kaufmann, London, 1988.

[252] Eugene Shikhovtsev. Biographical sketch of Hugh Everett, III. https://space.mit.edu/home/tegmark/everett/everett.html, 2003.

[253] P. W. Shor. Polynomial-Time Algorithms for Prime Factorization and Discrete Logarithms on a Quantum Computer. *ArXiv Quantum Physics e-prints*, August 1995.

[254] P.W. Shor. Algorithms for quantum computation: discrete logarithms and factoring. In *Proceedings 35th Annual Symposium on Foundations of Computer Science*, pages 124–134, Nov 1994.

[255] Herbert A. Simon. *The Science of the Artificial*. The MIT Press, 1969.

[256] Herbert A. Simon. *Models of my Life*. Basic Books, New York, 1991.

[257] Christine A. Skarda and Walter J. Freeman. How brains make chaos in order to make sense of the world. *Behavioral and Brain Sciences*, 10:161–195, 1987.

[258] S.M. Solaiman. Legal personality of robots, corporations, idols and chimpanzees: A quest for legitimacy. *Artificial Intelligence and Law*, 25:155–179, 2017.

[259] F. T. Sommer and A. Wichert. *Exploratory Analysis and Data Modeling in Functional Neuroimaging*. The MIT Press, 2002.

[260] Henry P. Stapp. *Mind, Matter and Quantum Mechanics, 3rd ed.* Springer, 2009.

[261] Z. Steinman, R. M.and Pizlo and F. J. Pizlo. Phi is not beta, and why wertheimer's discovery launched the gestalt revolution. *Vision Research.*, 40(17):2257–2264, 2000.

[262] P. Sterling. Principles of allostasis: optimal design, predictive regulation, pathophysiology and rational therapeutics. In J. Schulkin, editor, *Allostasis, Homeostasis, and the Costs of Adaptation*, pages 17–64. Cambridge: University Press, 2004.

[263] Tom Stonier. *Information and the Internal Structure of Universe.* Springer-Verlag, 1990.

[264] David G. Stork. Scientist on the set: An interview with Marvin Minsky. In David G. Stork, editor, *HAL's Legacy 2001's Computer as Dream and Reality*, chapter 2. The MIT Press, 1997.

[265] Leonard Susskind. *Quantum Mechanics Theoretical Minimum.* Penguin, 2015.

[266] Leó Szilárd. Über die entropieverminderung in einem thermodynamischen system bei eingriffen intelligenter wesen. *Zeitschrift für Physik*, 53:840–856, 1929.

[267] Luís Tarrataca and Andreas Wichert. Tree search and quantum computation. *Quantum Information Processing*, 10(4):475–500, 2011. 10.1007/s11128-010-0212-z.

[268] Luís Tarrataca and Andreas Wichert. Iterative quantum tree search. CiE 2012 - How the World Computes, 2012, 2012.

[269] Luís Tarrataca and Andreas Wichert. Quantum iterative deepening with an application to the halting problem. *PLOS ONE*, 8(3), 2013.

[270] Alfred Tarski. The semantic conception of truth and foundations of semantics. *Philos. and Phenom. Res.*, 4:241–376, 1944.

[271] Alfred Tarski. *Logic, Semantics,Metamathematics.* Oxford University Press, London, 1956.

[272] Alfred Tarski. *Pisma logiczno-filozoficzne. Prawda*, volume 1. Wydawnictwo Naukowe PWN, Warszawa, 1995.

[273] Nuo Tay, Chu Loo, and Mitja Perus. Face recognition with quantum associative networks using overcomplete gabor wavelet. *Cognitive Computation*, pages 1–6, 2010. 10.1007/s12559-010-9047-2.

[274] C. C. W. Taylor. *The Atomists: Leucippus and Democritus: Fragments (Phoenix Presocractic Series).* University of Toronto Press, 2010.

[275] Max Tegmark. The interpretation of quantum mechanics: Many worlds or many words? *ArXiv Quantum Physics e-prints*, abs/9709032, 1997.

[276] Richard F. Thompson. *The Brain: A Neuroscience Primer*. W H Freeman and Co, 1993.

[277] Flemming Topsoe. *Informationstheorie*. Teubner Sudienbucher, 1974.

[278] Daniel J. Treier and Walter A. Elwell. *Evangelical Dictionary of Theology*. Baker Academic, 2017.

[279] Jennifer S Trueblood, James M Yearsley, and Emmanuel M Pothos. A quantum probability framework for human probabilistic inference. *Journal of Experimental Psychology: General*, 146(9):1307, 2017.

[280] Elishai Ezra Tsur. *Neuromorphic Engineering*. CRC Pres, 2021.

[281] A. M. Turing. Computing machinery and intelligence. *Mind*, 59, 1950.

[282] A.M. Turing. On computable numbers, with an application to the entscheidungsproblem. In *Proceedings of the London Mathematical Society*, volume 2, pages 260–265, 1936.

[283] Lieven M. K. Vandersypen, Matthias Steffen, Gregory Breyta, Costantino S. Yannoni, Mark H. Sherwood, and Isaac L. Chuang. Experimental realization of shor's quantum factoring algorithm using nuclear magnetic resonance. *Nature*, 414:883–887, December 2001.

[284] Vlatko Vedral. Living in a quantum world. *Scientific American*, 304(6):38–43, 2011.

[285] E. von Holst and H. Mittelstaedt. Das reafferenzprinzip (wechselwirkungen zwischen zentralnervensystem und peripherie. *Naturwissenschaften*, 37:464–476, 1950.

[286] John von Neumann. First draft of a report on the edvac. Technical report, University of Pennsylvania, June 1945.

[287] David Wallace. Everettian rationality: defending deutsch's approach to probability in the everett interpretation. *Studies in History and Philosophy of Modern Physics*, 24:415–439, 2003.

[288] David Wallace. Quantum probability from subjective likelihood: Improving on deutsch's proof of the probability rule. *Studies in History and Philosophy of Modern Physics*, 38:311–332, 2007.

[289] David Wallece. *The Emergent Multiverse: Quantum Theory According to the Everett Interpretation*. Oxford University Press, 2010.

[290] R. T. Wallis. *Neoplatonism (Hackett Classics) 2nd Edition*. Hackett Publishing Company, 1995.

[291] H.G. Wells. *A Short History Of The World*. Cassell & Co, 1922.

[292] Paul Werbos. *Beyond Regression: New Tools for Prediction and Analysis in the Behavioral Sciences*. PhD thesis, Harvard University, 1974.

[293] Alfred North Whitehead and Bertrand Russell. *Principia Mathematica*. Andesite Press, 2015.

[294] A. Wichert. Associative computer: a hybrid connectionistic production system. *Cognitive Systems Research*, 6(2):111–144, 2005.

[295] A. Wichert, Birgit Abler, Jo Grothe, Henrik Walter, and F. T. Sommer. Exploratory analysis of event-related fmri demonstrated in a working memory study. In F.T. Sommer and A. Wichert, editors, *Exploratory Analysis and Data Modeling in Functional Neuroimaging*, chapter 5, pages 77–108. MIT Press, Boston, MA, 2002.

[296] A. Wichert and H. A. Kestler. A categorical based reanimation expert system. In E.C. Ifeachor, A. Sperduti, and A. Starita, editors, *Neural Networks and Expert Systems in Medicine and Healthcare*. World Scientific, 1998.

[297] A. Wichert, J. D. Pereira, and P. Carreira. Visual search light model for mental problem solving. *Neurocomputing*, 71(13-15):2806–2822, 2008.

[298] Andreas Wichert. A categorical expert system "jurassic". *Expert Systems with Application*, 19(3):149–158, 2000.

[299] Andreas Wichert. Associative diagnose. *Expert Systems*, 22(1):26–39, 2005.

[300] Andreas Wichert. Cell assemblies for diagnostic problem-solving. *Neurocomputing*, 69(7–9):810–824, 2006.

[301] Andreas Wichert. Sub-symbols and icons. *Cognitive Computation*, 1(4):342–347, 2009.

[302] Andreas Wichert. *Principles of Quantum Artificial Intelligence*. World Scientific, 2013.

[303] Andreas WIchert. Artificial intelligence and a universal quantum computer. *AI Communications*, 29(4):537–543, 2016.

[304] Andreas Wichert. Brain and mind in everett many-worlds. *Journal of Consciousness Exploration and Research,*, 7(19):817–822, 2016.

[305] Andreas Wichert. *Principles of Quantum Artificial Intelligence: Quantum Problem Solving and Machine Learning, 2nd Edition*. World Scientific, 2020.

[306] Andreas Wichert and Catarina Moreira. Probabilities and shannon's entropy in the everett many-worlds theory. *Frontiers in Physics: Interdisciplinary Physics*, 4(47), 2016.

[307] Andreas Wichert and Catarina Moreira. Balanced quantum-like model for decision making. In Springer, editor, *Proceedings of 11th International Conference on Quantum Interaction, Nice, France*, pages 79–90, September 3–5, 2018.

[308] Andreas Wichert, Catarina Moreira, and Peter Bruza. Quantum-like bayesian networks. *Entropy*, 22(2):170, 2020.

[309] Andreas Wichert and Luis Sa-Couto. *Machine Learning - A Journey to Deep Learning*. World Scientific, 2021.

[310] Andrzej Wichert. Pictorial reasoning with cell assemblies. *Connection Science*, 13(1), 2001.

[311] Norbert Wiener. *Cybernetics or Control and Communication in the Animal and the Machine, Reissue of the 1961 second edition*. The MIT Press, 2019.

[312] David E. Wilkins. "that's something I could not allow to happen": HAL and planning. In David G. Strok, editor, *HAL's Legacy 2001's Computer as Dream and Reality*, chapter 14, pages 305–331. The MIT Press, 1997.

[313] Colin P. Williams and Scott H. Clearwatter. *Explorations in Quantum Computing*. Springer-Verlag, 1997.

[314] D.J. Willshaw, O.P. Buneman, and H.C. Longuet-Higgins. Nonholgraphic associative memory. *Nature*, 222:960–962, 1969.

[315] Partick Henry Winston. *Artificial Intelligence*. Addison-Wesley, third edition, 1992.

[316] Peter Wittek. *Quantum Machine Learning, What Quantum Computing Means to Data Mining*. Elsevier Insights. Academic Press, 2014.

[317] Ludwig Wittgenstein. *Philosophical Investigations*. Wiley-Blackwell, 2009.

[318] Stephen Wolfram. *A New Kind of Science*. Wolfram Media, 2002.

[319] Yong Yao and Walter J. Freeman. Model of biological pattern recognition with spatially chaotic dynamics. *Neural Networks*, 3(2):153–170, 1990.

[320] Antonio Zadra and Robert Stickgold. *When Brains Dream: Understanding the Science and Mystery of Our Dreaming Minds*. W. W. Norton and Company, 2022.

[321] Arthur Zajonc. *Meditation as Contemplative Inquiry: When Knowing Becomes Love*. Lindisfarne Books, 2008.

[322] B. Zhou, E. Pöppel, and Y. Bao. In the jungle of time: the concept of identity as a way out. *Frontiers in Psychology*, 5:844, 2014.

Index